Random Operator Theory

Random Operator Theory

Reza Saadati
Department of Mathematics
Iran University of Science and Technology
Tehran, Iran

AMSTERDAM · BOSTON · HEIDELBERG · LONDON
NEW YORK · OXFORD · PARIS · SAN DIEGO
SAN FRANCISCO · SINGAPORE · SYDNEY · TOKYO
Academic Press is an imprint of Elsevier

Academic Press is an imprint of Elsevier
32 Jamestown Road, London NW1 7BY, UK
525 B Street, Suite 1800, San Diego, CA 92101-4495, USA
50 Hampshire Street, 5th Floor, Cambridge, MA 02139, USA
The Boulevard, Langford Lane, Kidlington, Oxford OX5 1GB, UK

British Library Cataloguing-in-Publication Data
A catalogue record for this book is available from the British Library

Library of Congress Cataloging-in-Publication Data
A catalog record for this book is available from the Library of Congress

ISBN: 978-0-12-805346-1

For information on all Academic Press publications
visit our website at https://www.elsevier.com/

Working together
to grow libraries in
developing countries

www.elsevier.com • www.bookaid.org

Publisher: Nikki Levy
Acquisition Editor: Glyn Jones
Editorial Project Manager: Anna Valutkevich
Production Project Manager: Poulouse Joseph
Designer: Mark Rogers

Typeset by SPi Global, India

To my beloved family

Preface

The notion of random normed space goes back to Sherstnev [46] and [18, 19, 45], who were dulled from Karl Menger [34], Bertold Schweizer and Abe Sklar [45] works. Some authors [5, 6, 29, 30, 31, 43] considered some properties of probabilistic normed spaces. Fixed point theory [14, 19], approximation theory, and stability of functional equations [4, 37, 38, 39] are studied at random normed space and its depended space, that is fuzzy normed space.

This book is introduced as a survey and research of the latest and new results on the following topics:

(i)   Basic theory of triangular norms
(ii) Topological structure of random normed spaces and fundamental theorems
(iii) Random Banach spaces
(iv) Random Banach algebras and fixed point theorem

Tehran, April 2016   Reza Saadati

Acknowledgment

It is my pleasure to acknowledge the superb assistance provided by the staff of Elsevier for the publication of the book.

Reza Saadati

Contents

Chapter 1
Preliminaries

In this chapter, we recall some definitions and results which will be used later on in the book.

1.1 Triangular norms

Triangular norms first appeared in the framework of probabilistic metric spaces in the work of Menger [34]. It turns also out that this is an essential operation in several fields. Triangular norms are an indispensable tool for the interpretation of the conjunction in fuzzy logics [23] and, subsequently, for the intersection of fuzzy sets [47]. They are, however, interesting mathematical objects for themselves. We refer to some papers and books for further details (see [24, 25, 26, 27], [18] and [45]).

Definition 1.1.1 A *triangular norm* (shortly, *t-norm*) is a binary operation on the unit interval $[0,1]$, i.e., a function $T : [0,1] \times [0,1] \to [0,1]$ such that, for all $a, b, c \in [0,1]$, the following four axioms are satisfied:

(T1) $T(a,b) = T(b,a)$ (commutativity);
(T2) $T(a,(T(b,c))) = T(T(a,b),c)$ (associativity);
(T3) $T(a,1) = a$ (boundary condition);
(T4) $T(a,b) \leq T(a,c)$ whenever $b \leq c$ (monotonicity).

The commutativity of (T1), the boundary condition (T3), and the monotonicity (T4) imply that, for each t-norm T and $x \in [0,1]$, the following boundary conditions are also satisfied:

$$T(x,1) = T(1,x) = x,$$

$$T(x,0) = T(0,x) = 0,$$

Random Operator Theory
http://dx.doi.org/10.1016/B978-0-12-805346-1.50001-9
Copyright © 2016 Elsevier Ltd. All rights reserved.

and so all the t-norms coincide with the boundary of the unit square $[0,1]^2$.

The monotonicity of a t-norm T in the second component (T4) is, together with the commutativity (T1), equivalent to the (joint) monotonicity in both components, that is,

$$T(x_1, y_1) \leq T(x_2, y_2) \qquad\qquad (1.1.1)$$

whenever $x_1 \leq x_2$ and $y_1 \leq y_2$.

Basic examples are the Lukasiewicz t-norm T_L:

$$T_L(a, b) = \max\{a + b - 1, 0\}$$

for all $a, b \in [0, 1]$ and the t-norms T_P, T_M, T_D are defined as follows:

$$T_P(a, b) := ab,$$
$$T_M(a, b) := \min\{a, b\},$$
$$T_D(a, b) := \begin{cases} \min\{a, b\}, & \text{if } \max\{a,b\}=1, \\ 0, & \text{otherwise.} \end{cases}$$

If, for any two t-norms T_1 and T_2, the inequality $T_1(x, y) \leq T_2(x, y)$ holds for all $(x, y) \in [0, 1]^2$, then we say that T_1 is *weaker* than T_2 or, equivalently, T_2 is *stronger* than T_1.

From (1.1.1), it follows that, for all $(x, y) \in [0, 1]^2$,

$$T(x, y) \leq T(x, 1) = x,$$

$$T(x, y) \leq T(1, y) = y.$$

Since $T(x, y) \geq 0 = T_D(x, y)$ for all $(x, y) \in (0, 1)^2$ holds trivially, for any t-norm T, we have

$$T_D \leq T \leq T_M,$$

that is, T_D is weaker and T_M is stronger than any other t-norms. Also, since $T_L < T_P$, we obtain the following ordering for four basic t-norms:

$$T_D < T_L < T_P < T_M.$$

Proposition 1.1.2 ([18])

(1)*The minimum T_M is the only t-norm satisfying $T(x, x) = x$ for all $x \in (0, 1)$.*

(2)*The weakest t-norm T_D is the only t-norm satisfying $T(x, x) = 0$ for all $x \in (0, 1)$.*

Proposition 1.1.3 ([18]) *A t-norm T is continuous if and only if it is continuous in its first component, i.e., for all $y \in [0, 1]$, if the one-place function*

$$T(\cdot, y) : [0, 1] \to [0, 1], \quad x \mapsto T(x, y),$$

is continuous.

For example, the minimum T_M and Łukasiewicz t-norm T_L are continuous, but the t-norm $T^\triangle$ defined by

$$T^\triangle(x, y) := \begin{cases} \frac{xy}{2}, & \text{if } \max\{x, y\} < 1, \\ xy, & \text{otherwise,} \end{cases}$$

for all $x, y \in [0, 1]$ is not continuous.

Definition 1.1.4

(1) A t-norm T is said to be *strictly monotone* if

$$T(x, y) < T(x, z)$$

whenever $x \in (0, 1)$ and $y < z$.

(2) A t-norm T is said to be *strict* if it is continuous and strictly monotone.

For example, the t-norm $T^\triangle$ is strictly monotone, but the minimum T_M and Łukasiewicz t-norm T_L are not strictly monotone.

Proposition 1.1.5 ([18]) *A t-norm T is strictly monotone if and only if*

$$T(x, y) = T(x, z), \ x > 0 \quad \Longrightarrow \quad y = z.$$

If T is a t-norm, then $x_T^{(n)}$ for all $x \in [0, 1]$ and $n \geq 0$ is defined by 1 if $n = 0$ and $T(x_T^{(n-1)}, x)$ if $n \geq 1$.

Definition 1.1.6 A t-norm T is said to be *Archimedean* if, for all $(x, y) \in (0, 1)^2$, there exists an integer $n \geq 1$ such that

$$x_T^{(n)} < y.$$

Proposition 1.1.7 ([18]) *A t-norm T is Archimedean if and only if, for all $x \in (0, 1)$,*

$$\lim_{n \to \infty} x_T^{(n)} = 0.$$

Proposition 1.1.8 ([18]) *If t-norm T is Archimedean, then, for all $x \in (0, 1)$, we have*

$$T(x, x) < x.$$

For example, the product T_p, Łukasiewicz t-norm T_L and the weakest t-norm T_D are all Archimedean, but the minimum T_M is not an Archimedean t-norm.

A t-norm T is said to be *of Hadžić-type* (denoted by $T \in \mathcal{H}$) if the family $\{x_T^{(n)}\}$ is equicontinuous at $x = 1$, that is, for any $\varepsilon \in (0,1)$, there exists $\delta \in (0,1)$ such that

$$x > 1 - \delta \quad \Longrightarrow \quad x_T^{(n)} > 1 - \varepsilon \tag{1.1.2}$$

for all $n \geq 1$.

The t-norm T_M is a trivial example of Hadžić type, but T_P is not of Hadžić type.

Proposition 1.1.9 ([18]) *If a continuous t-norm T is Archimedean, then it cannot be a t-norm of Hadžić-type.*

Other important t-norms are as follows (see [20]):

(1)The *Sugeno-Weber family* $\{T_\lambda^{SW}\}_{\lambda \in [-1,\infty]}$ is defined by $T_{-1}^{SW} = T_D$, $T_\infty^{SW} = T_P$ and

$$T_\lambda^{SW}(x,y) = \max\left\{0, \frac{x + y - 1 + \lambda xy}{1 + \lambda}\right\}$$

if $\lambda \in (-1, \infty)$.

(2)The *Domby family* $\{T_\lambda^D\}_{\lambda \in [0,\infty]}$ is defined by T_D, if $\lambda = 0$, T_M if $\lambda = \infty$ and

$$T_\lambda^D(x,y) = \frac{1}{1 + ((\frac{1-x}{x})^\lambda + (\frac{1-y}{y})^\lambda)^{1/\lambda}}$$

if $\lambda \in (0, \infty)$.

(3)The *Aczel-Alsina family* $\{T_\lambda^{AA}\}_{\lambda \in [0,\infty]}$ is defined by T_D, if $\lambda = 0$, T_M if $\lambda = \infty$ and

$$T_\lambda^{AA}(x,y) = e^{-(|\log x|^\lambda + |\log y|^\lambda)^{1/\lambda}}$$

if $\lambda \in (0, \infty)$.

A t-norm T can be extended (by associativity) in a unique way to an n-array operation taking, for any $(x_1, \cdots, x_n) \in [0,1]^n$, the value $T(x_1, \cdots, x_n)$ defined by

$$\mathrm{T}_{i=1}^0 x_i = 1, \quad \mathrm{T}_{i=1}^n x_i = T(\mathrm{T}_{i=1}^{n-1} x_i, x_n) = T(x_1, \cdots, x_n).$$

The t-norm T can also be extended to a countable operation taking, for any sequence $\{x_n\}$ in $[0,1]$, the value

$$\mathrm{T}_{i=1}^\infty x_i = \lim_{n \to \infty} \mathrm{T}_{i=1}^n x_i. \tag{1.1.3}$$

The limit on the right side of (1.1.3) exists since the sequence $\{T_{i=1}^n x_i\}$ is nonincreasing and bounded from below.

Proposition 1.1.10 ([20])

(1)*For $T \geq T_L$ the following implication holds:*

$$\lim_{n \to \infty} T_{i=1}^\infty x_{n+i} = 1 \quad \Longleftrightarrow \quad \sum_{n=1}^\infty (1 - x_n) < \infty.$$

(2)*If T is of Hadžić-type, then we have*

$$\lim_{n \to \infty} T_{i=1}^\infty x_{n+i} = 1$$

for any sequence $\{x_n\}_{n \geq 1}$ in $[0, 1]$ such that $\lim_{n \to \infty} x_n = 1$.

(3)*If $T \in \{T_\lambda^{AA}\}_{\lambda \in (0,\infty)} \cup \{T_\lambda^D\}_{\lambda \in (0,\infty)}$, then we have*

$$\lim_{n \to \infty} T_{i=1}^\infty x_{n+i} = 1 \quad \Longleftrightarrow \quad \sum_{n=1}^\infty (1 - x_n)^\alpha < \infty.$$

(4)*If $T \in \{T_\lambda^{SW}\}_{\lambda \in [-1,\infty)}$, then we have*

$$\lim_{n \to \infty} T_{i=1}^\infty x_{n+i} = 1 \quad \Longleftrightarrow \quad \sum_{n=1}^\infty (1 - x_n) < \infty.$$

Definition 1.1.11 Let T and T' be two continuous t-norms. Then, we say that T' *dominates* T (denoted by $T' \gg T$) if, for all $x_1, x_2, y_1, y_2 \in [0, 1]$,

$$T[T'(x_1, x_2), T'(y_1, y_2)] \leq T'[T(x_1, y_1), T(x_2, y_2)].$$

We say the t-norm T has Σ property and write $T \in \Sigma$ whenever, for any $\lambda \in (0, 1)$, there exists $\gamma \in (0, 1)$ (which does not depend on n) such that

$$T^{n-1}(1 - \gamma, \cdots, 1 - \gamma) > 1 - \lambda \tag{1.1.4}$$

for each $n \geq 1$.

1.2 Distribution functions

Let Δ^+ denotes the space of all distribution functions, that is, the space of all mappings $F : \mathbb{R} \cup \{-\infty, +\infty\} \longrightarrow [0, 1]$ such that F is left-continuous, non-decreasing on $\mathbb{R}$, $F(0) = 0$, and $F(+\infty) = 1$. D^+ is a subset of Δ^+ consisting of all functions $F \in \Delta^+$ for which $l^- F(+\infty) = 1$, where $l^- f(x)$ denotes the left limit of the function f at the point x, that is, $l^- f(x) = \lim_{t \to x^-} f(t)$. The space Δ^+ is partially ordered by the usual point-wise ordering of functions, that is, $F \leq G$ if and only if $F(t) \leq G(t)$ for all $t \in \mathbb{R}$. The maximal element for Δ^+ in this order is the distribution function ε_0 given by

$$\varepsilon_0(t) = \begin{cases} 0, & \text{if } t \leq 0, \\ 1, & \text{if } t > 0. \end{cases}$$

Example 1.2.1 The function $G(t)$ defined by

$$G(t) = \begin{cases} 0, & \text{if } t \leq 0, \\ 1 - e^{-t}, & \text{if } t > 0, \end{cases}$$

is a distribution function. Since $\lim_{t \to \infty} G(t) = 1$, $G \in D^+$. Note that $G(t + s) \geq T_p(G(t), G(s))$ for each $t, s > 0$.

Example 1.2.2 The function $F(t)$ defined by

$$F(t) = \begin{cases} 0, & \text{if } t \leq 0, \\ t, & \text{if } 0 \leq t \leq 1, \\ 1, & \text{if } 1 \leq t, \end{cases}$$

is a distribution function. Since $\lim_{t \to \infty} F(t) = 1$, $F \in D^+$. Note that $F(t + s) \geq T_M(F(t), F(s))$ for all $t, s > 0$.

Example 1.2.3 ([5]) The function $G_p(t)$ defined by

$$G_p(t) = \begin{cases} 0, & \text{if } t \leq 0, \\ \exp(-|p|^{1/2}), & \text{if } 0 < t < +\infty, \\ 1, & \text{if } t = +\infty, \end{cases}$$

is a distribution function. Since $\lim_{t \to \infty} G_p(t) \neq 1$, $G \in \Delta^+ \setminus D^+$. Note that $G_p(t + s) \geq T_M(G_p(t), G_p(s))$ for all $t, s > 0$.

Definition 1.2.4 A *nonmeasure distribution function* is a function $\nu : \mathbb{R} \to [0, 1]$ which is right continuous on $\mathbb{R}$, nonincreasing and $\inf_{t \in \mathbb{R}} \nu(t) = 0$, $\sup_{t \in \mathbb{R}} \nu(t) = 1$.

We denote by B the family of all nonmeasure distribution functions and by G a special element of B defined by

$$G(t) = \begin{cases} 1, & \text{if } t \le 0, \\ 0, & \text{if } t > 0. \end{cases}$$

If X is a nonempty set, then $\nu : X \longrightarrow B$ is called a *probabilistic nonmeasure* on X and $\nu(x)$ is denoted by ν_x.

Let $\mathcal{L} = (L, \ge_L)$ be a complete lattice, that is, a partially ordered set in which every nonempty subset admits supremum, infimum and $0_{\mathcal{L}} = \inf L$, $1_{\mathcal{L}} = \sup L$. The space of latticetic random distribution functions, denoted by Δ_L^+, is defined as the set of all mappings $F : \mathbb{R} \cup \{-\infty, +\infty\} \to L$ such that F is left-continuous and nondecreasing on $\mathbb{R}$, $F(0) = 0_{\mathcal{L}}$, and $F(+\infty) = 1_{\mathcal{L}}$.

$D_L^+ \subseteq \Delta_L^+$ is defined as $D_L^+ = \{F \in \Delta_L^+ : l^- F(+\infty) = 1_{\mathcal{L}}\}$, where $l^- f(x)$ denotes the left limit of the function f at the point x. The space Δ_L^+ is partially ordered by the usual point-wise ordering of functions, that is, $F \ge G$ if and only if $F(t) \ge_L G(t)$ for all $t \in \mathbb{R}$. The maximal element for Δ_L^+ in this order is the distribution function given by

$$\varepsilon_0(t) = \begin{cases} 0_{\mathcal{L}}, & \text{if } t \le 0, \\ 1_{\mathcal{L}}, & \text{if } t > 0. \end{cases}$$

Chapter 2
Random Banach Spaces

In this chapter, we present some facts on the random Banach spaces and their properties.

2.1 Random normed spaces

Random (probabilistic) normed spaces were introduced by Šerstnev in 1962 [46] by means of a definition that was closely modelled on the theory of (classical) normed spaces and used to study the problem of best approximation in statistics. In the sequel, we shall adopt usual terminology, notation, and conventions of the theory of random normed spaces, as in [5, 6, 31, 45].

Definition 2.1.1 A *Menger probabilistic metric space* (or *random metric spaces*) is a triple $(X, \mathcal{F}, T)$, where X is a nonempty set, T is a continuous t-norm, and $\mathcal{F}$ is a mapping from $X \times X$ into D^+ such that, if $F_{x,y}$ denotes the value of $\mathcal{F}$ at a point $(x, y) \in X \times X$, the following conditions hold: for all x, y, z in X,

(PM1) $F_{x,y}(t) = \varepsilon_0(t)$ for all $t > 0$ if and only if $x = y$;
(PM2) $F_{x,y}(t) = F_{y,x}(t)$;
(PM3) $F_{x,z}(t + s) \geq T(F_{x,y}(t), F_{y,z}(s))$ for all $x, y, z \in X$ and $t, s \geq 0$.

Definition 2.1.2 ([46]) A *random normed space* (briefly, a RN-space) or *a Šerstnev (Sherstnev) probabilistic normed space* (briefly, a Šerstnev PN-space) is a triple (X, μ, T), where X is a vector space, T is a continuous t-norm, and μ is a mapping from X into D^+ such that the following conditions hold:

(RN1) $\mu_x(t) = \varepsilon_0(t)$ for all $t > 0$ if and only if $x = 0$ (0 is the null vector in X);
(RN2) $\mu_{\alpha x}(t) = \mu_x\left(\frac{t}{|\alpha|}\right)$ for all $x \in X$ and $\alpha \neq 0$;

Random Operator Theory
http://dx.doi.org/10.1016/B978-0-12-805346-1.50002-0

(RN3)　$\mu_{x+y}(t+s) \geq T(\mu_x(t), \mu_y(s))$ for all $x, y \in X$ and $t, s \geq 0$, where μ_x denotes the value of μ at a point $x \in X$.

Note that a *triangular function* $\tau : \Delta^+ \times \Delta^+ \to \Delta^+$ is a binary operation on Δ^+ which is associative, commutative, and nondecreasing in each argument and has ε_0 as the unit, that is, for all $F, G, H \in \Delta^+$,

$$\begin{aligned}
\tau(\tau(F,G), H) &= \tau(F, \tau(G, H)), \\
\tau(F, G) &= \tau(G, F), \\
\tau(F, \varepsilon_0) &= F, \\
F \leq G &\Longrightarrow \tau(F, H) \leq \tau(G, H).
\end{aligned}$$

The continuity of a triangular function means the continuity with respect to the topology of weak convergence in Δ^+. Triangular functions are recursively defined by $\tau^1 = \tau$ and

$$\tau^n(F_1, \cdots, F_{n+1}) = \tau(\tau^{n-1}(F_1, \cdots, F_n), F_{n+1})$$

for each $n \geq 2$.

Typical continuous triangular functions are as follows:

$$\tau_T(F, G)(x) = \sup_{s+t=x} T(F(s), G(t)),$$

and

$$\tau_{T^*}(F, G) = \inf_{s+t=x} T^*(F(s), G(t)),$$

where T is a continuous t-norm, that is, a continuous binary operation on $[0, 1]$ that is commutative, associative, nondecreasing in each variable, and has 1 as the identity element and T^* is a continuous t-conorm, that is, a continuous binary operation on $[0, 1]$ which is related to the continuous t-norm T through $T^*(x, y) = 1 - T(1 - x, 1 - y)$.

Examples of such t-norms and t-conorms are M and M^*, respectively, defined by

$$M(x, y) = \min(x, y)$$

and

$$M^*(x, y) = \max(x, y).$$

Let τ_1 and τ_2 be two triangular functions. Then τ_1 dominates τ_2 (which is denoted by $\tau_1 \gg \tau_2$) if, for all $F_1, F_2, G_1, G_2 \in \Delta^+$,

$$\tau_1(\tau_2(F_1, G_1), \tau_2(F_2, G_2)) \geq \tau_2(\tau_1(F_1, F_2), \tau_1(G_1, G_2)).$$

In 1993, Alsina, Schweizer, and Sklar gave a new definition of a probabilistic normed space [5] as follows:

A *probabilistic normed space* (briefly, PN-space) is a quadruple (V, ν, τ, τ^*), where V is a real vector space, τ, τ^* are continuous triangle functions, and ν is a mapping from $V \to \Delta^+$ such that, for all $p, q \in V$, the following conditions hold:

(PN1) $\nu_p = \varepsilon_0$ if and only if $p = \theta$, where θ is the null vector in V;
(PN2) $\nu_{-p} = \nu_p$ for all $p \in V$;
(PN3) $\nu_{p+q} \geq \tau(\nu_p, \nu_q)$ for all $p, q \in V$;
(PN4) $\nu_p \leq \tau^*(\nu_{\alpha p}, \nu_{(1-\alpha)p})$ for all $\alpha \in [0, 1]$.

If the inequality (PN4) is replaced by the equality $\nu_p = \tau_M(\nu_{\alpha p}, \nu_{(1-\alpha)p})$, then the PN-space (V, ν, τ, τ^*) is called a *Šerstnev probabilistic normed space* or a *random normed space* (see Definition 2.1.2) and, as a consequence, we have the following condition stronger than (N2):

$$\nu_{\lambda p}(x) = \nu_p\left(\frac{x}{|\lambda|}\right)$$

for all $p \in V$, $\lambda \neq 0$ and $x \in \mathbb{R}$.

Example 2.1.3 Let $(X, \|\cdot\|)$ be a linear normed spaces. Define a mapping

$$\mu_x(t) = \begin{cases} 0, & \text{if } t \leq 0, \\ \frac{t}{t+\|x\|}, & \text{if } t > 0. \end{cases}$$

Then (X, μ, T_p) is a random normed space. In fact, (RN1) and (RN2) are obvious. Now, we show (RN3).

$$
\begin{aligned}
T_p((\mu_x(t), \mu_y(s))) &= \frac{t}{t + \|x\|} \cdot \frac{s}{s + \|y\|} \\
&= \frac{1}{1 + \frac{\|x\|}{t}} \cdot \frac{1}{1 + \frac{\|y\|}{s}} \\
&\leq \frac{1}{1 + \frac{\|x\|}{t+s}} \cdot \frac{1}{1 + \frac{\|y\|}{t+s}} \\
&\leq \frac{1}{1 + \frac{\|x\|+\|y\|}{t+s}} \\
&\leq \frac{1}{1 + \frac{\|x+y\|}{t+s}} \\
&= \frac{t+s}{t + s + \|x + y\|} \\
&= \mu_{x+y}(t + s)
\end{aligned}
$$

for all $x, y \in X$ and $t, s \geq 0$. Also, (X, μ, T_M) is a random normed space.

Example 2.1.4 Let $(X, \| \cdot \|)$ be a linear normed space. Define a mapping

$$\mu_x(t) = \begin{cases} 0, & \text{if } t \leq 0, \\ e^{-\left(\frac{\|x\|}{t}\right)}, & \text{if } t > 0. \end{cases}$$

Then (X, μ, T_p) is a random normed space. In fact, (RN1) and (RN2) are obvious and so, now, we show (RN3).

$$\begin{aligned} T_p((\mu_x(t), \mu_y(s)) &= e^{-\left(\frac{\|x\|}{t}\right)} \cdot e^{-\left(\frac{\|y\|}{s}\right)} \\ &\leq e^{-\left(\frac{\|x\|}{t+s}\right)} \cdot e^{-\left(\frac{\|y\|}{t+s}\right)} \\ &= e^{-\left(\frac{\|x\|+\|y\|}{t+s}\right)} \\ &\leq e^{-\left(\frac{\|x+y\|}{t+s}\right)} \\ &= \mu_{x+y}(t+s) \end{aligned}$$

for all $x, y \in X$ and $t, s \geq 0$. Also, (X, μ, T_M) is a random normed space.

Example 2.1.5 ([38]) Let $(X, \| \cdot \|)$ be a linear normed space. For all $x \in X$, define a mapping

$$\mu_x(t) = \begin{cases} \max\{1 - \frac{\|x\|}{t}, 0\}, & \text{if } t > 0, \\ 0, & \text{if } t \leq 0. \end{cases}$$

Then, (X, μ, T_L) is an RN-space (this was essentially proved by Musthari in [40], see also [42]). Indeed, we have

$$\mu_x(t) = 1 \quad \Longrightarrow \quad \frac{\|x\|}{t} = 0 \quad \longrightarrow \quad x = 0$$

for all $t > 0$ and, obviously,

$$\mu_{\lambda x}(t) = \mu_x\left(\frac{t}{\lambda}\right)$$

for all $x \in X$ and $t > 0$. Next, for any $x, y \in X$ and $t, s > 0$, we have

$$\begin{aligned} \mu_{x+y}(t+s) &= \max\left\{1 - \frac{\|x+y\|}{t+s}, 0\right\} \\ &= \max\left\{1 - \left\|\frac{x+y}{t+s}\right\|, 0\right\} \\ &= \max\left\{1 - \left\|\frac{x}{t+s} + \frac{y}{t+s}\right\|, 0\right\} \\ &\geq \max\left\{1 - \left\|\frac{x}{t}\right\| - \left\|\frac{y}{s}\right\|, 0\right\} \\ &= T_L(\mu_x(t), \mu_y(s)). \end{aligned}$$

2.2 Random topological structures

In this section, we review some topological structures of random normed spaces, for more details please see [2, 15].

Definition 2.2.1 Let (X, μ, T) be an RN-space. We define the *open ball* $B_x(r, t)$ and the closed ball $B_x[r, t]$ with center $x \in X$ and radius $0 < r < 1$ for all $t > 0$ as follows:

$$B_x(r, t) = \{y \in X \ : \ \mu_{x-y}(t) > 1 - r\},$$

$$B_x[r, t] = \{y \in X \ : \ \mu_{x-y}(t) \geq 1 - r\},$$

respectively.

Theorem 2.2.2 *Let (X, μ, T) be an RN-space. Every open ball $B_x(r, t)$ is an open set.*

Proof. Let $B_x(r, t)$ be an open ball with center x and radius r for all $t > 0$. Let $y \in B_x(r, t)$. Then $\mu_{x-y}(t) > 1 - r$. Since $\mu_{x-y}(t) > 1 - r$, there exists $t_0 \in (0, t)$ such that $\mu_{x-y}(t_0) > 1 - r$. Put $r_0 = \mu_{x,y}(t_0)$. Since $r_0 > 1 - r$, there exists $s \in (0, 1)$ such that $r_0 > 1 - s > 1 - r$. Now, for any r_0 and s such that $r_0 > 1 - s$, there exists $r_1 \in (0, 1)$ such that $T(r_0, r_1) > 1 - s$. Consider the open ball $B_y(1 - r_1, t - t_0)$.

Now, we claim that $B_y(1 - r_1, t - t_0) \subset B_x(r, t)$. In fact, let $z \in B_y(1 - r_1, t - t_0)$. Then $\mu_{y-z}(t - t_0) > r_1$ and so

$$\begin{aligned} \mu_{x-z}(t) &\geq T(\mu_{x-y}(t_0), \mu_{y-z}(t - t_0)) \\ &\geq T(r_0, r_1) \\ &\geq 1 - s \\ &> 1 - r. \end{aligned}$$

Thus $z \in B_x(r, t)$ and hence $B_y(1 - r_1, t - t_0) \subset B_x(r, t)$. This completes the proof.

Now, different kinds of topologies can be introduced in a random normed space [45]. The (r, t)-*topology* is introduced by a family of neighborhoods

$$\{B_x(r, t)\}_{x \in X, t > 0, r \in (0,1)}.$$

In fact, every random norm μ on X generates a topology $((r, t)$-topology) on X which has a base as the family of open sets of the form

$$\{B_x(r, t)\}_{x \in X, t > 0, r \in (0,1)}.$$

Remark 2.2.3 Since $\left\{ B_x \left(\frac{1}{n}, \frac{1}{n} \right) : n \geq 1 \right\}$ is a local base at x, the (r, t)-topology is first countable.

Theorem 2.2.4 *Every RN-space (X, μ, T) is a* Hausdorff *space.*

Proof. Let (X, μ, T) be an RN-space. Let x and y be two distinct points in X and $t > 0$. Then $0 < \mu_{x-y}(t) < 1$. Put $r = \mu_{x-y}(t)$. For each $r_0 \in (r, 1)$, there exists r_1 such that $T(r_1, r_1) \geq r_0$. Consider the open balls $B_x(1 - r_1, \frac{t}{2})$ and $B_y(1 - r_1, \frac{t}{2})$. Then, clearly, $B_x(1 - r_1, \frac{t}{2}) \cap B_y(1 - r_1, \frac{t}{2}) = \emptyset$. In fact, if there exists

$$z \in B_x \left(1 - r_1, \frac{t}{2} \right) \cap B_y \left(1 - r_1, \frac{t}{2} \right),$$

then we have

$$
\begin{aligned}
r &= \mu_{x-y}(t) \\
&\geq T \left(\mu_{x-z} \left(\frac{t}{2} \right), \mu_{y-z} \left(\frac{t}{2} \right) \right) \\
&\geq T(r_1, r_1) \\
&\geq r_0 \\
&> r,
\end{aligned}
$$

which is a contradiction. Hence, (X, μ, T) is a Hausdorff space. This completes the proof.

Definition 2.2.5 Let (X, μ, T) be an RN-space. A subset A of X is said to be *R-bounded* if there exist $t > 0$ and $r \in (0, 1)$ such that $\mu_{x-y}(t) > 1 - r$ for all $x, y \in A$.

Theorem 2.2.6 *Every compact subset A of an RN-space (X, μ, T) is R-bounded.*

Proof. Let A be a *compact* subset of an RN-space (X, μ, T). Fix $t > 0$, $0 < r < 1$ and consider an open cover $\{ B_x(r, t) : x \in A \}$. Since A is compact, there exist $x_1, x_2, \cdots, x_n \in A$ such that

$$A \subseteq \bigcup_{i=1}^{n} B_{x_i}(r, t).$$

Let $x, y \in A$. Then $x \in B_{x_i}(r, t)$ and $y \in B_{x_j}(r, t)$ for some $i, j \geq 1$. Thus, we have $\mu_{x-x_i}(t) > 1 - r$ and $\mu_{y-x_j}(t) > 1 - r$. Now, let

$$\alpha = \min \{ \mu_{x_i, x_j}(t) : 1 \leq i, j \leq n \}.$$

Then, we have $\alpha > 0$ and

$$\mu_{x-y}(3t) \geq T^2(\mu_{x-x_i}(t), \mu_{x_i,x_j}(t), \mu_{y-x_j}(t))$$
$$\geq T^2(1-r, 1-r, \alpha)$$
$$> 1-s.$$

Taking $t' = 3t$, it follows that $\mu_{x-y}(t') > 1 - s$ for all $x, y \in A$. Hence, A is R-bounded. This completes the proof.

Remark 2.2.7 In an RN-space (X, μ, T), every compact set is closed and R-bounded.

Definition 2.2.8 Let (X, μ, T) be an RN-space.

(1) A sequence $\{x_n\}$ in X is said to be *convergent* to a point $x \in X$ if, for any $\epsilon > 0$ and $\lambda > 0$, there exists a positive integer N such that

$$\mu_{x_n-x}(\epsilon) > 1 - \lambda$$

whenever $n \geq N$.

(2) A sequence $\{x_n\}$ in X is called a *Cauchy sequence* if, for any $\epsilon > 0$ and $\lambda > 0$, there exists a positive integer N such that

$$\mu_{x_n-x_m}(\epsilon) > 1 - \lambda$$

whenever $n \geq m \geq N$.

(3) An RN-space (X, μ, T) is said to be *complete* if every Cauchy sequence in X is convergent to a point in X.

Theorem 2.2.9 ([45]) *If (X, μ, T) is an RN-space and $\{x_n\}$ is a sequence such that $x_n \to x$, then $\lim_{n\to\infty} \mu_{x_n}(t) = \mu_x(t)$ almost everywhere.*

Theorem 2.2.10 *Let (X, μ, T) be an RN-space such that every Cauchy sequence in X has a convergent subsequence. Then (X, μ, T) is complete.*

Proof. Let $\{x_n\}$ be a Cauchy sequence in X and $\{x_{i_n}\}$ be a subsequence of $\{x_n\}$ which converges to a point $x \in X$.

Now, we prove that $x_n \to x$. Let $t > 0$ and $\epsilon \in (0, 1)$ such that

$$T(1-r, 1-r) \geq 1 - \epsilon.$$

Since $\{x_n\}$ is a Cauchy sequence, there exists $n_0 \geq 1$ such that

$$\mu_{x_m-x_n}(t) > 1 - r$$

for all $m, n \geq n_0$. Since $x_{i_n} \to x$, there exists a positive integer i_p such that $i_p > n_0$ and

$$\mu_{x_{i_p}-x}\left(\frac{t}{2}\right) > 1 - r.$$

Then, if $n \geq n_0$, we have

$$\mu_{x_n-x}(t) \geq T\left(\mu_{x_n-x_{i_p}}\left(\frac{t}{2}\right), \mu_{x_{i_p}-x}\left(\frac{t}{2}\right)\right)$$
$$> T(1-r, 1-r)$$
$$\geq 1 - \epsilon.$$

Therefore, $x_n \to x$ and hence, (X, μ, T) is complete. This completes the proof.

Lemma 2.2.11 *Let* (X, μ, T) *be an RN-space. If we define*

$$F_{x,y}(t) = \mu_{x-y}(t)$$

for all $x, y \in X$ *and* $t > 0$, *then* F *is a random (probabilistic) metric on* X, *which is called the random (probabilistic) metric induced by the random norm* μ.

Lemma 2.2.12 *A random (probabilistic) metric* F *which is induced by a random norm on an RN-space* (X, μ, T) *has the following properties: for all* $x, y, z \in X$ *and scalar* $\alpha \neq 0$,

(1)$F_{x+z,y+z}(t) = F_{x,y}(t)$;
(2)$F_{\alpha x, \alpha y}(t) = F_{x,y}\left(\frac{t}{|\alpha|}\right)$.

Proof. We have the following:

$$F_{x+z,y+z}(t) = \mu_{(x+z)-(y+z)}(t) = \mu_{x-y}(t) = F_{x,y}(t)$$

and, also,

$$F_{\alpha x, \alpha y}(t) = \mu_{\alpha x - \alpha y}(t) = \mu_{x-y}\left(\frac{t}{|\alpha|}\right) = F_{x,y}\left(\frac{t}{|\alpha|}\right).$$

Therefore, we have (1) and (2). This completes the proof.

Lemma 2.2.13 *If* (X, μ, T) *is an RN-space, then we have*

(1)*The function* $(x, y) \longrightarrow x + y$ *is continuous.*
(2)*The function* $(\alpha, x) \longrightarrow \alpha x$ *is continuous.*

Proof. If $x_n \longrightarrow x$ and $y_n \longrightarrow y$ as $n \longrightarrow \infty$, then we have

$$\mu_{(x_n+y_n)-(x+y)}(t) \geq T\left(\mu_{x_n-x}\left(\frac{t}{2}\right), \mu_{y_n-y}\left(\frac{t}{2}\right)\right) \longrightarrow 1$$

as $n \longrightarrow \infty$. This proves (1).

Now, if $x_n \longrightarrow x$ and $\alpha_n \longrightarrow \alpha$ as $n \longrightarrow \infty$, where $\alpha_n \neq 0$, then we have

$$\mu_{\alpha_n x_n - \alpha x}(t) = \mu_{\alpha_n(x_n - x) + x(\alpha_n - \alpha)}(t)$$

$$\geq T\left(\mu_{\alpha_n(x_n-x)}\left(\frac{t}{2}\right)\mu_{x(\alpha_n-\alpha)}\left(\frac{t}{2}\right)\right)$$

$$= T\left(\mu_{x_n-x}\left(\frac{t}{2\alpha_n}\right), \mu_x\left(\frac{t}{2(\alpha_n - \alpha)}\right)\right) \longrightarrow 1$$

as $n \longrightarrow \infty$. This proves (2). This completes the proof.

Definition 2.2.14 An RN-space (X, μ, T) is called a *random Banach space* whenever X is complete with respect to the random metric induced by random norm.

Lemma 2.2.15 *Let (X, μ, T) be an RN-space and define*

$$E_{\lambda,\mu} : X \longrightarrow \mathbb{R}^+ \cup \{0\}$$

by

$$E_{\lambda,\mu}(x) = \inf\{t > 0 : \mu_x(t) > 1 - \lambda\}$$

for all $\lambda \in (0,1)$ and $x \in X$. Then we have

(1)$E_{\lambda,\mu}(\alpha x) = |\alpha| E_{\lambda,\mu}(x)$ *for all $x \in X$ and $\alpha \in \mathbb{R}$;*

(2)*If T has Σ property, then, for any $\alpha \in (0,1)$, there exists $\beta \in (0,1)$ such that*

$$E_{\gamma,\mu}(x_1 + \cdots + x_n) \leq E_{\lambda,\mu}(x_1) + \cdots + E_{\lambda,\mu}(x_n)$$

for all $x, y \in X$;

(3)*A sequence $\{x_n\}$ is convergent with respect to the random norm μ if and only if $E_{\lambda,\mu}(x_n - x) \to 0$. Also, the sequence $\{x_n\}$ is a Cauchy sequence with respect to the random norm μ if and only if it is a Cauchy sequence with $E_{\lambda,\mu}$.*

Proof. For (1), we have

$$E_{\lambda,\mu}(\alpha x) = \inf\{t > 0 : \mu_{\alpha x}(t) > 1 - \lambda\}$$

$$= \inf\left\{t > 0 : \mu_x\left(\tfrac{t}{|\alpha|}\right) > 1 - \lambda\right\}$$

$$= |\alpha| \inf\{t > 0 : \mu_x(t) > 1 - \lambda\}$$

$$= |\alpha| E_{\lambda,\mu}(x).$$

For (2), by Σ property, for all $\alpha \in (0,1)$, we can find $\lambda \in (0,1)$ such that

$$T^{n-1}(1 - \lambda, \cdots, 1 - \lambda) \geq 1 - \alpha.$$

Thus we have

$$\mu_{x_1+\cdots+x_n}(E_{\lambda,\mu}(x_1) + \cdots + E_{\lambda,\mu}(x_n) + n\delta)$$
$$\geq T^{n-1}(\mu_{x_1}(E_{\lambda,\mathcal{M}}(x_1) + \delta), \cdots, \mu_{x_n}(E_{\lambda,\mathcal{P}}(x_n) + \delta))$$
$$\geq T(1 - \lambda, \cdots, 1 - \lambda)$$
$$\geq 1 - \alpha$$

for all $\delta > 0$, which implies that

$$E_{\alpha,\mu}(x_1 + \cdots + x_n) \leq E_{\lambda,\mu}(x_1) + \cdots + E_{\lambda,\mu}(x_n) + n\delta.$$

Since $\delta > 0$ is arbitrary, we have

$$E_{\alpha,\mu}(x_1 + \cdots + x_n) \leq E_{\lambda,\mu}(x_1) + \cdots + E_{\lambda,\mu}(x_n).$$

For (3), since μ is continuous, $E_{\lambda,\mu}(x)$ is not an element of the set $\{t > 0 : \mu_x(t) > 1 - \lambda\}$ for all $x \in X$ with $x \neq 0$. Hence, we have

$$\mu_{x_n-x}(\eta) > 1 - \lambda \iff E_{\lambda,\mu}(x_n - x) < \eta$$

for all $\eta > 0$. This completes the proof.

Definition 2.2.16 A function f from an RN-space (X, μ, T) to an RN-space (Y, ν, T') is said to be *uniformly continuous* if, for all $r \in (0,1)$ and $t > 0$, there exist $r_0 \in (0,1)$ and $t_0 > 0$ such that

$$\mu_{x-y}(t_0) > 1 - r_0 \implies \nu_{f(x),f(y)}(t) > 1 - r.$$

Theorem 2.2.17 (Uniform Continuity Theorem) *If f is a continuous function from a compact RN-space (X, μ, T) to an RN-space (Y, ν, T'), then f is uniformly continuous.*

Proof. Let $s \in (0,1)$ and $t > 0$ be given. Then we can find $r \in (0,1)$ such that

$$T'(1 - r, 1 - r) > 1 - s.$$

Since $f : X \to Y$ is continuous, for any $x \in X$, we can find $r_x \in (0,1)$ and $t_x > 0$ such that

$$\mu_{x-y}(t_x) > 1 - r_x \implies \nu_{f(x)-f(y)}\left(\frac{t}{2}\right) > 1 - r.$$

But $r_x \in (0,1)$ and then we can find $s_x < r_x$ such that

$$T(1 - s_x, 1 - s_x) > 1 - r_x.$$

Since X is compact and

$$\left\{ B_x \left(s_x, \frac{t_x}{2} \right) : x \in X \right\}$$

is an open covering of X, there exist $x_1, x_2, \cdots, x_k$ in X such that

$$X = \bigcup_{i=1}^{k} B_{x_i} \left(s_{x_i}, \frac{t_{x_i}}{2} \right).$$

Put $s_0 = \min s_{x_i}$ and $t_0 = \min \frac{t_{x_i}}{2}$, $i = 1, 2, \cdots, k$. For any $x, y \in X$, if $\mu_{x-y}(t_0) > 1 - s_0$, then $\mu_{x-y} \left(\frac{t_{x_i}}{2} \right) > 1 - s_{x_i}$. Since $x \in X$, there exists $x_i \in X$ such that

$$\mu_{x-x_i} \left(\frac{t_{x_i}}{2} \right) > 1 - s_{x_i}.$$

Hence we have

$$\nu_{f(x),f(x_i)} \left(\frac{t}{2} \right) > 1 - r.$$

Now, note that

$$\begin{aligned}
\mu_{y-x_i}(t_{x_i}) &\geq T \left(\mu_{x-y} \left(\frac{t_{x_i}}{2} \right), \mu_{x-x_i} \left(\frac{t_{x_i}}{2} \right) \right) \\
&\geq T(1 - s_{x_i}, 1 - s_{x_i}) \\
&> 1 - r_{x_i}.
\end{aligned}$$

Therefore, we have

$$\nu_{f(y)-f(x_i)} \left(\frac{t}{2} \right) > 1 - r$$

and so

$$\begin{aligned}
\nu_{f(x)-f(y)}(t) &\geq T \left(\nu_{f(x)-f(x_i)} \left(\tfrac{t}{2} \right), \nu_{f(y)-f(x_i)} \left(\tfrac{t}{2} \right) \right) \\
&\geq T(1 - r, 1 - r) \\
&> 1 - s.
\end{aligned}$$

Therefore, f is uniformly continuous. This completes the proof.

Remark 2.2.18 Let f be an uniformly continuous function from an RN-space (X, μ, T) to an RN-space (Y, ν, T'). If $\{x_n\}$ is a Cauchy sequence in X, then $\{f(x_n)\}$ is also a Cauchy sequence in Y.

Theorem 2.2.19 *Every compact RN-space is separable.*

Proof. Let (X, μ, T) be a compact RN-space. Let $r \in (0, 1)$ and $t > 0$. Since X is compact, there exist $x_1, x_2, \cdots, x_n$ in X such that

$$X = \bigcup_{i=1}^{n} B_{x_i}(r,t).$$

In particular, for each $n \geq 1$, we can choose a finite subset A_n of X such that

$$X = \bigcup_{a \in A_n} B_a\left(r_n, \frac{1}{n}\right)$$

in which $r_n \in (0,1)$. Let

$$A = \bigcup_{n \geq 1} A_n.$$

Then A is countable.

Now, we claim that $X \subset \overline{A}$. Let $x \in X$. Then, for each $n \geq 1$, there exists $a_n \in A_n$ such that $x \in B_{a_n}(r_n, \frac{1}{n})$. Thus $\{a_n\}$ converges to the point $x \in X$. But, since $a_n \in A$ for all $n \geq 1$, $x \in \overline{A}$ and so A is dense in X. Therefore, X is separable. This completes the proof.

Definition 2.2.20 Let X be a nonempty set and (Y, ν, T') be an RN-space. Then a sequence $\{f_n\}$ of functions from X to Y is said to *converge uniformly* to a function f from X to Y if, for any $r \in (0,1)$ and $t > 0$, there exists $n_0 \geq 1$ such that

$$\nu_{f_n(x)-f(x)}(t) > 1 - r$$

for all $n \geq n_0$ and $x \in X$.

Definition 2.2.21 A family $\mathcal{F}$ of functions from an RN-space (X, μ, T) to a complete RN-space (Y, ν, T') is said to be *equicontinuous* if, for any $r \in (0,1)$ and $t > 0$, there exist $r_0 \subset (0,1)$ and $t_0 > 0$ such that

$$\mu_{x-y}(t_0) > 1 - r_0 \quad \Longrightarrow \quad \nu_{f(x)-f(y)}(t) > 1 - r$$

for all $f \in \mathcal{F}$.

Lemma 2.2.22 *Let $\{f_n\}$ be an equicontinuous sequence of functions from an RN-space (X, μ, T) to a complete RN-space (Y, ν, T'). If $\{f_n\}$ converges for each point of a dense subset D of X, then $\{f_n\}$ converges for each point of X and the limit function is continuous.*

Proof. Let $s \in (0,1)$ and $t > 0$ be given. Then, we can find $r \in (0,1)$ such that

$$T'^2(1-r, 1-r, 1-r) > 1 - s.$$

Since $\mathcal{F} = \{f_n\}$ is an equicontinuous family, for any $r \in (0,1)$ and $t > 0$, there exist $r_1 \in (0,1)$ and $t_1 > 1$ such that, for each $x, y \in X$,

$$\mu_{x-y}(t_1) > 1 - r_1 \quad \implies \quad \nu_{f_n(x)-f_n(y)}\left(\frac{t}{3}\right) > 1 - r$$

for all $f_n \in \mathcal{F}$. Since D is dense in X, there exists

$$y \in B_x(r_1, t_1) \bigcap D$$

and $\{f_n(y)\}$ converges for the point y. Since $\{f_n(y)\}$ is a Cauchy sequence, for any $r \in (0,1)$ and $t > 0$, there exists $n_0 \geq 1$ such that

$$\nu_{f_n(y)-f_m(y)}\left(\frac{t}{3}\right) > 1 - r$$

for all $m, n \geq n_0$. Now, for any $x \in X$, we have

$$
\begin{aligned}
&\nu_{f_n(x)-f_m(x)}(t) \\
&\geq T'^2\left(\nu_{f_n(x)-f_n(y)}\left(\tfrac{t}{3}\right), \nu_{f_n(y)-f_m(y)}\left(\tfrac{t}{3}\right), \nu_{f_m(x)-f_m(y)}\left(\tfrac{t}{3}\right)\right) \\
&\geq T'^2(1-r, 1-r, 1-r) \\
&> 1 - s.
\end{aligned}
$$

Hence $\{f_n(x)\}$ is a Cauchy sequence in Y. Since Y is complete, $f_n(x)$ converges and so let $f(x) = \lim f_n(x)$.

Now, we claim that f is continuous. Let $s_o \in 1 - r$ and $t_0 > 0$ be given. Then, we can find $r_0 \in 1 - r$ such that

$$T'^2(1-r_0, 1-r_0, 1-r_0) > 1 - s_0.$$

Since $\mathcal{F}$ is equicontinuous, for any $r_0 \in (0,1)$ and $t_0 > 0$, there exist $r_2 \in (0,1)$ and $t_2 > 0$ such that

$$\mu_{x-y}(t_2) > 1 - r_2 \quad \implies \quad \nu_{f_n(x)-f_n(y)}\left(\frac{t_0}{3}\right) > 1 - r_0$$

for all $f_n \in \mathcal{F}$. Since $f_n(x)$ converges to $f(x)$, for any $r_0 \in (0,1)$ and $t_0 > 0$, there exists $n_1 \geq 1$ such that

$$\nu_{f_n(x)-f(x)}\left(\frac{t_0}{3}\right) > 1 - r_0.$$

Also, since $f_n(y)$ converges to $f(y)$, for any $r_0 \in (0,1)$ and $t_0 > 0$, there exists $n_2 \geq 1$ such that

$$\nu_{f_n(y)-f(y)}\left(\frac{t_0}{3}\right) > 1 - r_0$$

for all $n \geq n_2$. Now, for all $n \geq \max\{n_1, n_2\}$, we have

$$\nu_{f(x)-f(y)}(t_0)$$
$$\geq T'^2 \left(\nu_{f(x)-f_n(x)} \left(\tfrac{t_0}{3} \right), \nu_{f_n(x)-f_n(y)} \left(\tfrac{t_0}{3} \right), \nu_{f_n(y)-f(y)} \left(\tfrac{t_0}{3} \right) \right)$$
$$\geq T'^2 (1 - r_0, 1 - r_0, 1 - r_0)$$
$$> 1 - s_0.$$

Therefore, f is continuous. This completes the proof.

Theorem 2.2.23 (Ascoli-Arzela Theorem) *Let (X, μ, T) be a compact RN-space and (Y, ν, T') a complete RN-space. Let $\mathcal{F}$ be an equicontinuous family of functions from X to Y. If $\{f_n\}$ is a sequence in $\mathcal{F}$ such that*

$$\overline{\{f_n(x) : n \in \mathbb{N}\}}$$

is a compact subset of Y for any $x \in X$, then there exists a continuous function f from X to Y and a subsequence $\{g_n\}$ of $\{f_n\}$ such that $\{g_n\}$ converges uniformly to f on X.

Proof. Since (X, μ, T) is a compact RN-space, by Theorem 2.2.19, X is separable. Let

$$D = \{x_i : i = 1, 2, \cdots\}$$

be a countable dense subset of X. By hypothesis, for each $i \geq 1$,

$$\overline{\{f_n(x_i) : n \geq 1\}}$$

is a compact subset of Y. Since every $\mathcal{L}$-fuzzy metric space is first countable space, every compact subset of Y is sequentially compact. Thus, by standard argument, we have a subsequence $\{g_n\}$ of $\{f_n\}$ such that $\{g_n(x_i)\}$ converges for each $i \geq 1$. Thus, by Lemma 2.2.22, there exists a continuous function f from X to Y such that $\{g_n(x)\}$ converges to $f(x)$ for all $x \in X$.

Now, we claim that $\{g_n\}$ converges uniformly to a function f on X. Let $s \in (0, 1)$ and $t > 0$ be given. Then, we can find $r \in (0, 1)$ such that

$$T'^2 (1 - r, 1 - r, 1 - r) > 1 - s.$$

Since $\mathcal{F}$ is equicontinuous, there exist $r_1 \in (0, 1)$ and $t_1 > 0$ such that

$$\mu_{x-y}(t_1) > 1 - r_1 \quad \Longrightarrow \quad \nu_{g_n(x), g_n(y)} \left(\frac{t}{3} \right) > 1 - r$$

for all $n \geq 1$. Since X is compact, by Theorem 2.2.17, f is uniformly continuous. Hence, for any $r \in (0, 1)$ and $t > 0$, there exist $r_2 \in (0, 1)$ and $t_2 > 0$ such that

$$\mu_{x-y}(t_2) > 1 - r_2 \quad \Longrightarrow \quad \nu_{f(x)-f(y)} \left(\frac{t}{3} \right) > 1 - r$$

for all $x, y \in X$. Let $r_0 = \min\{r_1, r_2\}$ and $t_0 = \min\{t_1, t_2\}$. Since X is compact and D is dense in X, we have

$$X = \bigcup_{i=1}^{k} B_{x_i}(r_0, t_0)$$

for some $k \geq 1$. Thus, for any $x \in X$, there exists i, $i \leq i \leq k$, such that

$$\mu_{x - x_i}(t_0) > 1 - r_0.$$

But, since $r_0 = \min\{r_1, r_2\}$ and $t_0 = \min\{t_1, t_2\}$, we have, by the equicontinuity of $\mathcal{F}$,

$$\nu_{g_n(x) - g_n(x_i)}\left(\frac{t}{3}\right) > 1 - r$$

and we also have, by the uniform continuity of f,

$$\nu_{f(x) - f(x_i)}\left(\frac{t}{3}\right) > 1 - r.$$

Since $\{g_n(x_j)\}$ converges to $f(x_j)$, for any $r \in (0, 1)$ and $t > 0$, there exists $n_0 \geq 1$ such that

$$\nu_{g_n(x_j) - f(x_j)}\left(\frac{t}{3}\right) > 1 - r$$

for all $n \geq n_0$. Now, for all $x \in X$, we have

$$
\begin{aligned}
&\nu_{g_n(x) - f(x)}(t) \\
&\geq T'^2\left(\nu_{g_n(x) - g_n(x_i)}\left(\tfrac{t}{3}\right), \nu_{g_n(x_i) - f(x_i)}\left(\tfrac{t}{3}\right), \nu_{f(x_i) - f(x)}\left(\tfrac{t}{3}\right)\right) \\
&\geq T'^2(1 - r, 1 - r, 1 - r) \\
&> 1 - s.
\end{aligned}
$$

Therefore, $\{g_n\}$ converges uniformly to a function f on X. This completes the proof.

We recall that a subset A is said to be R-bounded in (X, μ, T), if there exist $t_0 > 0$ and $r_0 \in (0, 1)$ such that $\mu_x(t_0) > 1 - r_0$ for all $x \in A$.

Lemma 2.2.24 *A subset A of $\mathbb{R}$ is R-bounded in $(\mathbb{R}, \mu, T)$ if and only if it is bounded in $\mathbb{R}$.*

Proof. Let A be a subset in $\mathbb{R}$ which is R-bounded in $(\mathbb{R}, \mu, T)$. Then there exist $t_0 > 0$ and $r_0 \in (0, 1)$ such that $\mu_x(t_0) > 1 - r_0$ for all $x \in A$. Thus, we have

$$t_0 \geq E_{r_0, \mu}(x) = |x| E_{r_0, \mu}(1).$$

Now, $E_{r_0, \mu}(1) \neq 0$. If we put $k = \frac{t_0}{E_{r_0, \mu}(1)}$, then we have $|x| \leq k$ for all $x \in A$, that is, A is bounded in $\mathbb{R}$.

The converse is easy to see. This completes the proof.

Lemma 2.2.25 *A sequence $\{\beta_n\}$ is convergent in an RN-space $(\mathbb{R}, \mu, T)$ if and only if it is convergent in $(\mathbb{R}, |\cdot|)$.*

Proof. Let $\beta_n \longrightarrow \beta$ in $\mathbb{R}$. Then, by Lemma 2.2.15 (1), we have

$$E_{\lambda,\mu}(\beta_n - \beta) = |\beta_n - \beta| E_{\lambda,\mu}(1) \longrightarrow 0.$$

Thus, by Lemma 2.2.15 (3), $\beta_n \xrightarrow{\mu} \beta$.

Conversely, let $\beta_n \xrightarrow{\mu} \beta$. Then, by Lemma 2.2.15,

$$\lim_{n \to +\infty} |\beta_n - \beta| E_{\lambda,\mu}(1) = \lim_{n \to +\infty} E_{\lambda,\mu}(\beta_n - \beta) = 0.$$

Now, $E_{\lambda,\mu}(1) \neq 0$ and so $\beta_n \longrightarrow \beta$ in $\mathbb{R}$. This completes the proof.

Corollary 2.2.26 *If a real sequence $\{\beta_n\}$ is R-bounded, then it has at least one limit point.*

Lemma 2.2.27 *A subset A of $\mathbb{R}$ is R-bounded in $(\mathbb{R}, \mu, T)$ if and only if it is bounded in $\mathbb{R}$.*

Proof. Let the subset A is R-bounded in $(\mathbb{R}, \mu, T)$. Then there exist $t_0 > 0$ and $r_0 \in (0,1)$ such that

$$\mu_x(t_0) > 1 - r_0$$

for all $x \in A$ and so

$$t_0 \geq E_{r_0,\mu}(x) = |x| E_{r_0,\mu}(1).$$

Now, $E_{r_0,\mu}(1) \neq 0$. If we put $k = \frac{t_0}{E_{r_0,\mu}(1)}$, then we have $|x| \leq k$ for all $x \in A$, that is, A is bounded in $\mathbb{R}$.

The converse is easy. This completes the proof.

Definition 2.2.28 A triple $(\mathbb{R}^n, \Phi, T)$ is called a *random Euclidean normed space* if T is a continuous t-norm and $\Phi_x(t)$ is a random Euclidean norm defined by

$$\Phi_x(t) = \prod_{j=1}^{n} \mu_{x_j}(t),$$

where $\prod_{j=1}^{n} a_j = T'^{n-1}(a_1, \cdots, a_n)$, $T' \gg T$, $x = (x_1, \cdots, x_n)$, $t > 0$, and μ is a random norm.

For example, let $\Phi_x(t) = \exp\left(\frac{\|x\|}{t}\right)^{-1}$, $\mu_{x_j}(t) = \exp\left(\frac{|x_j|}{t}\right)^{-1}$, and $T = \min$. Then, we have $\Phi_x(t) = \min_j \mu_{x_j}(t)$ or, equivalently, $\|x\| = \max_j |x_j|$.

Lemma 2.2.29 *Suppose that the hypotheses of Definition 2.2.28 are satisfied. Then $(\mathbb{R}^n, \Phi, T)$ is an RN-space.*

Proof. The properties of (RN1) and (RN2) follow immediately from the definition. For the triangle inequality (RN3), suppose that $x, y \in X$ and $t, s > 0$. Then, we have

$$
\begin{aligned}
T(\Phi_x(t), \Phi_y(s)) &= T\Big(\prod_{j=1}^n \mathcal{P}_{x_j}(t), \prod_{j=1}^n \mathcal{P}_{y_j}(s)\Big) \\
&= T(T'^{n-1}(\mathcal{P}_{x_1}(t), \cdots, \mathcal{P}_{x_n}(t)), T'^{n-1}(\mathcal{P}_{y_1}(t), \cdots, \mathcal{P}_{y_n}(t)) \\
&\leq T'^{n-1}(T(\mathcal{P}_{x_1}(t), \mathcal{P}_{y_1}(t)), \cdots, T(\mathcal{P}_{x_n}(t), \mathcal{P}_{y_n}(t)) \\
&\leq T'^{n-1}(\mathcal{P}_{x_1+y_1}(t+s), \cdots, \mathcal{P}_{x_n+y_n}(t+s)) \\
&= \prod_{j=1}^n \mathcal{P}_{x_j+y_j}(t+s) \\
&= \Phi_{x+y}(t+s).
\end{aligned}
$$

This completes the proof.

Lemma 2.2.30 *Suppose that $(\mathbb{R}^n, \Phi, T)$ is a random Euclidean normed space and A is an infinite and R-bounded subset of $\mathbb{R}^n$. Then A has at least one limit point.*

Proof. Let $\{x^{(m)}\}$ be an infinite sequence in A. Since A is R-bounded, so is $\{x^{(m)}\}_{m \geq 1}$. Therefore, there exist $t_0 > 0$ and $r_0 \in (0, 1)$ such that

$$1 - r_0 < \Phi_x(t_0)$$

for all $x \in A$, which implies that $E_{r_0, \Phi}(x) \leq t_0$. However, we have

$$
\begin{aligned}
E_{r_0, \Phi}(x) &= \inf\{t > 0 : 1 - r_0 < \Phi_x(t)\} \\
&= \inf\Big\{t > 0 : 1 - r_0 < \prod_{j=1}^n \mu_{x_j}(t)\Big\} \\
&\geq \inf\{t > 0 : 1 - r_0 < \mu_{x_j}(t)\} \\
&= E_{r_0, \mu}(x_j)
\end{aligned}
$$

for each $1 \leq j \leq n$. Therefore, $|x_j| \leq k$ in which $k = \frac{t_0}{E_{r_0, \mu}(1)}$, that is, the real sequences $\{x_j^{(m)}\}$ for each $j \in \{1, \cdots, n\}$ are bounded. Hence, there exists a subsequence $\{x_1^{(m_{k_1})}\}$ which converges to x_1 in A with respect to the random norm μ. The corresponding sequence $\{x_2^{(m_{k_1})}\}$ is bounded and so there exists a subsequence $\{x_2^{(m_{k_2})}\}$ of $\{x_2^{(m_{k_1})}\}$ which converges to x_2 with respect to the random norm μ.

Continuing like this, we find a subsequence $\{x^{(m_k)}\}$ converging to $x = (x_1, \cdots, x_n) \in \mathbb{R}^n$. This completes the proof.

Lemma 2.2.31 *Let* $(\mathbb{R}^n, \Phi, T)$ *be a random Euclidean normed space. Let* $\{Q_1, Q_2, \cdots\}$ *be a countable collection of nonempty subsets in* $\mathbb{R}^n$ *such that* $Q_{k+1} \subseteq Q_k$, *each* Q_k *is closed and* Q_1 *is R-bounded. Then* $\cap_{k=1}^\infty Q_k$ *is nonempty and closed.*

Proof. Using the above lemma, the proof proceeds as in the classical case (see Theorem 3.25 in [7]).

We call an n-dimensional ball $B_x(r, t)$ a *rational ball* if $x \in \mathbb{Q}^n$, $r_0 \in (0, 1)$, and $t \in \mathbb{Q}^+$.

Theorem 2.2.32 *Let* $(\mathbb{R}^n, \Phi, T)$ *be a random Euclidean normed space in which* T *satisfies* Σ *property. Let* $G = \{A_1, A_2, \cdots\}$ *be a countable collection of n-dimensional rational open balls. If* $x \in \mathbb{R}^n$ *and* S *is an open subset of* $\mathbb{R}^n$ *containing* x, *then there exists* $A_k \in G$ *such that* $x \in A_k \subseteq S$ *for some* $k \geq 1$.

Proof. Since $x \in S$ and S is open, there exist $r \in (0, 1)$ and $t > 0$ such that $B_x(r, t) \subseteq S$. By Σ property, we can find $\eta \in (0, 1)$ such that $1 - r < T(1 - \eta, 1 - \eta)$. Let $\{\xi_k\}_{k=1}^n$ be a finite sequence such that $1 - \eta < \prod_{k=1}^n (1 - \xi_k)$ and $x = (x_1, \cdots, x_n)$. Then we can find $y = (y_1, \cdots, y_n) \in \mathbb{Q}^n$ such that $(1 - \xi_k) < \mu_{x_k - y_k}(\frac{t}{2})$. Therefore, we have

$$1 - \eta < \prod_{k=1}^n (1 - \xi_k) \leq \Phi_{x-y}\left(\frac{t}{2}\right) = \prod_{k=1}^n \mu_{x_k - y_k}\left(\frac{t}{2}\right)$$

and so $x \in B_y\left(\eta, \frac{t}{2}\right)$.

Now, we prove that $B_y\left(\eta, \frac{t}{2}\right) \subseteq B_x(r, t)$. Let $z \in B_y\left(\eta, \frac{t}{2}\right)$. Then $\Phi_{y-z}\left(\frac{t}{2}\right) > 1 - \eta$ and hence

$$1 - r < T(1 - \eta, 1 - \eta) \leq T\left(\Phi_{x-y}\left(\frac{t}{2}\right), \Phi_{y-z}\left(\frac{t}{2}\right)\right) \leq \Phi_{x-z}(t).$$

On the other hand, there exists $t_0 \in \mathbb{Q}$ such that $t_0 < \frac{t}{2}$ and $x \in B_y(\eta, t_0) \subseteq B_y\left(\eta, \frac{t}{2}\right) \subseteq B_x(r, t) \subseteq S$. Now, $B_y(\eta, t_0) \in G$. This completes the proof.

Corollary 2.2.33 *In a random Euclidean normed space* $(\mathbb{R}^n, \Phi, T)$ *in which* T *has* Σ *property, every closed and R-bounded set is compact.*

Proof. The proof is similar to the proof of Theorem 3.29 in [7].

Corollary 2.2.34 *Let* $(\mathbb{R}^n, \Phi, T)$ *be a random Euclidean normed space in which* T *has* Σ *property and* $S \subseteq \mathbb{R}^n$. *Then* S *is compact set if and only if it is R-bounded and closed.*

Corollary 2.2.35 *The random Euclidean normed space* $(\mathbb{R}^n, \Phi, T)$ *is complete.*

Proof. Let $\{x_m\}$ be a Cauchy sequence in the random Euclidean normed space $(\mathbb{R}^n, \Phi, T)$. Since

$$E_{\lambda, \Phi}(x_n - x_m) = \inf\{t > 0 : \Phi_{x_n - x_m}(t) > 1 - \lambda\}$$

$$= \inf\{t > 0 : \prod_{j=1}^{n} \mathcal{P}_{x_{m,j} - x_{n,j}}(t) > 1 - \lambda\}$$

$$\geq \inf\{t > 0 : \mathcal{P}_{x_{m,j} - x_{n,j}}(t) > 1 - \lambda\}$$

$$= E_{\lambda, \mathcal{P}}(x_{m,j} - x_{n,j}) = |x_{m,j} - x_{n,j}| E_{\lambda, \mathcal{P}}(1),$$

the sequence $\{x_{m,j}\}$ for each $j = 1, \cdots, n$ is a Cauchy sequence in $\mathbb{R}$ and so it is convergent to $x_j \in \mathbb{R}$. Then, by Lemma 2.2.15, the sequence $\{x_{m,j}\}$ is convergent in RN-space $(\mathbb{R}, \mu, T)$.

Now, we prove that $\{x_m\}$ is convergent to $x = (x_1, \cdots, x_n)$. In fact, we have

$$\lim_m \Phi_{x_m - x}(t) = \lim_m \prod_{j=1}^{n} \mathcal{P}_{x_{m,j} - x_j}(t) = T'^{n-1}(1, \cdots, 1) = 1.$$

This completes the proof.

2.3 Fundamental Theorems in Random Banach Spaces

In this section, we discuss some important results dealing with topological isomorphisms and also give the proofs of Open Mapping Theorem, Closed Graph Theorem, and some other fundamental theorems in the framework of random Banach spaces.

Theorem 2.3.1 Let $\{x_1, \cdots, x_n\}$ be a linearly independent set of vectors in vector space X and (X, μ, T) be an RN-space. Then there exist $c \neq 0$ and an RN-space $(\mathbb{R}, \mu', T)$ such that, for every choice of the n real scalars $\alpha_1, \cdots, \alpha_n$,

$$\mu_{\alpha_1 x_1 + \cdots + \alpha_n x_n}(t) \leq \mu'_{c \sum_{j=1}^{n} |\alpha_j|}(t). \tag{2.3.1}$$

Proof. Put $s = |\alpha_1| + \cdots + |\alpha_n|$. If $s = 0$, all α_j's must be zero and so (2.3.1) holds for any c. Let $s > 0$. Then (2.3.1) is equivalent to the inequality that we obtain from (2.3.1) by dividing by s and putting $\beta_j = \frac{\alpha_j}{s}$, that is,

$$\mu_{\beta_1 x_1 + \cdots + \beta_n x_n}(t') \leq \mu'_c(t'), \tag{2.3.2}$$

where $t' = \frac{t}{s}$ and $\sum_{j=1}^{n} |\beta_j| = 1$. Hence, it suffices to prove the existence of $c \neq 0$ and the random norm μ' such that (2.3.2) holds. Suppose that this is not true. Then there exists a sequence $\{y_m\}$ of vectors

$$y_m = \beta_{1,m} x_1 + \cdots + \beta_{n,m} x_n, \quad \sum_{j=1}^{n} |\beta_{j,m}| = 1,$$

such that

$$\mu_{y_m}(t) \longrightarrow 1$$

as $m \longrightarrow \infty$ for any $t > 0$. Since $\sum_{j=1}^{n} |\beta_{j,m}| = 1$, we have $|\beta_{j,m}| \leq 1$ and so, by Lemma 2.2.24, the sequence of $\{\beta_{j,m}\}$ is R-bounded. According to Corollary 2.2.26, $\{\beta_{1,m}\}$ has a convergent subsequence. Let β_1 denotes the limit of the subsequence and let $\{y_{1,m}\}$ denotes the corresponding subsequence of $\{y_m\}$. By the same argument, $\{y_{1,m}\}$ has a subsequence $\{y_{2,m}\}$, since $\{\beta_{2,m}\}$ has a convergent subsequence, let β_2 denotes the limit. Continuing this process, after n steps, we obtain a subsequence $\{y_{n,m}\}_{m \geq 1}$ of $\{y_m\}$ such that

$$y_{n,m} = \sum_{j=1}^{n} \gamma_{j,m} x_j,$$

where $\sum_{j=1}^{n} |\gamma_{j,m}| = 1$, and $\gamma_{j,m} \longrightarrow \beta_j$ as $m \longrightarrow \infty$. By Lemma 2.2.15 (2), for any $\alpha \in (0, 1)$, there exists $\lambda \in (0, 1)$ such that

$$E_{\alpha,\mu}\left(y_{n,m} - \sum_{j=1}^{n} \beta_j x_j\right) = E_{\alpha,\mu}\left(\sum_{j=1}^{n}(\gamma_{j,m} - \beta_j)x_j\right)$$

$$\leq \sum_{j=1}^{n} |\gamma_{j,m} - \beta_j| E_{\lambda,\mu}(x_j) \longrightarrow 0$$

as $m \longrightarrow \infty$. By Lemma 2.2.15 (3), we conclude

$$\lim_{m \to \infty} y_{n,m} = \sum_{j=1}^{n} \beta_j x_j,$$

where $\sum_{j=1}^{n} |\beta_j| = 1$, and so all β_j cannot be zero. Put $y = \sum_{j=1}^{n} \beta_j x_j$. Since $\{x_1, \cdots, x_n\}$ is a linearly independent set, we have $y \neq 0$. Since $\mu_{y_m}(t) \longrightarrow 1$, by the assumption, we have $\mu_{y_{n,m}}(t) \longrightarrow 1$. Hence, we have

$$\mu_y(t) = \mu_{(y - y_{n,m}) + y_{n,m}}(t)$$
$$\geq T\left(\mu_{y - y_{n,m}} t/2\right), \mu_{y_{n,m}}(t/2)\right) \longrightarrow 1$$

and so $y = 0$, which is a contradiction. This completes the proof.

Definition 2.3.2 Let (X, μ, T) and (X, ν, T') be two RN-spaces. Then two random norms μ and ν are said to be *equivalent* whenever $x_n \xrightarrow{\mu} x$ in (X, μ, T) if and only if $x_n \xrightarrow{\nu} x$ in (X, ν, T').

Theorem 2.3.3 *In a finite dimensional vector space X, every two random norms μ and ν are equivalent.*

Proof. Let $\dim X = n$ and $\{v_1, \cdots, v_n\}$ be a basis for X. Then every $x \in X$ has a unique representation $x = \sum_{j=1}^{n} \alpha_j v_j$. Let $x_m \xrightarrow{\mu} x$ in (X, μ, T), but, for each $m \geq 1$, suppose that x_m has a unique representation, that is,

$$x_m = \alpha_{1,m} v_1 + \cdots + \alpha_{n,m} v_n.$$

By Theorem 2.3.1, there exist $c \neq 0$ and the random norm μ' such that (2.3.1) holds. Thus, we have

$$\mu_{x_m - x}(t) \leq \mu'_{c \sum_{j=1}^{n} |\alpha_{j,m} - \alpha_j|}(t) \leq \mu'_{c|\alpha_{j,m} - \alpha_j|}(t).$$

Now, if $m \longrightarrow \infty$, then we have

$$\mu_{x_m - x}(t) \longrightarrow 1$$

for all $t > 0$ and hence $|\alpha_{j,m} - \alpha_j| \longrightarrow 0$ in $\mathbb{R}$.

On the other hand, by Lemma 2.2.15 (2), for any $\alpha \in (0, 1)$, there exists $\lambda \in (0, 1)$ such that

$$E_{\alpha,\nu}(x_m - x) \leq \sum_{j=1}^{n} |\alpha_{j,m} - \alpha_j| E_{\lambda,\nu}(v_j).$$

Since $|\alpha_{j,m} - \alpha_j| \longrightarrow 0$, we have $x_m \xrightarrow{\nu} x$ in (X, ν, T'). Therefore, with the same argument, $x_m \longrightarrow x$ in (X, ν, T') implies $x_m \longrightarrow x$ in (X, μ, T). This completes the proof.

Definition 2.3.4 A linear operator $\Lambda : (X, \mu, T) \longrightarrow (Y, \nu, T')$ is said to be *random bounded* if there exists a constant $h \in \mathbb{R} - \{0\}$ such that, for all $x \in X$ and $t > 0$,

$$\nu_{\Lambda x}(t) \geq \mu_{hx}(t). \qquad (2.3.3)$$

Note that, by Lemma 2.2.15 and the last definition, we have

$$
\begin{aligned}
E_{\lambda,\nu}(\Lambda x) &= \inf\{t > 0 : \nu_{\Lambda x}(t) > 1 - \lambda\} \\
&\leq \inf\{t > 0 : \mu_x(t/|h|) > 1 - \lambda\} \\
&= |h| \inf\{t > 0 : \mu_x(t) > 1 - \lambda\} \\
&= |h| E_{\lambda,\mu}(x).
\end{aligned}
$$

Theorem 2.3.5 *Every linear operator $\Lambda : (X, \mu, T) \longrightarrow (Y, \nu, T')$ is random bounded if and only if it is continuous.*

Proof. By (2.3.3), every random bounded linear operator is continuous.

Now, we prove the converse. Let the linear operator Λ be continuous but is not random bounded. Then, for each $n \geq 1$, there exists $x_n \in X$ such that $E_{\lambda,\nu}(\Lambda x_n) \geq n E_{\lambda,\mu}(p_n)$.

If we let

$$y_n = \frac{x_n}{n E_{\lambda,\mu}(x_n)},$$

then it is easy to see $y_n \to 0$, but $\{\Lambda y_n\}$ does not tend to 0. This completes the proof.

Definition 2.3.6 A linear operator $\Lambda : (X, \mu, T) \longrightarrow (Y, \nu, T')$ is a *random topological isomorphism* if Λ is one-to-one, onto, and both Λ, Λ^{-1} are continuous. The RN-spaces (X, μ, T) and (Y, ν, T') for which such a Λ exists are said to be *random topologically isomorphic*.

Lemma 2.3.7 *A linear operator $\Lambda : (X, \mu, T) \longrightarrow (Y, \nu, T')$ is random topological isomorphism if Λ is onto and there exist constants $a, b \neq 0$ such that*

$$\mu_{ax}(t) \leq \nu_{\Lambda x}(t) \leq \mu_{bx}(t).$$

Proof. By the hypothesis, Λ is random bounded and, by last theorem, is continuous. Since $\Lambda x = 0$ implies that

$$1 = \nu_{\Lambda x}(t) \leq \mu_x \left(\frac{t}{|b|} \right)$$

and so $x = 0$, it follows that Λ is one-to-one. Thus Λ^{-1} exists and, since

$$\nu_{\Lambda x}(t) \leq \mu_{bx}(t)$$

is equivalent to

$$\nu_y(t) \leq \mu_{b\Lambda^{-1}y}(t) = \mu_{\Lambda^{-1}y}\left(\frac{t}{|b|}\right)$$

or

$$\nu_{\frac{1}{b}y}(t) \leq \mu_{\Lambda^{-1}y}(t),$$

where $y = \Lambda x$, we see that Λ^{-1} is random bounded and, by last theorem, is continuous. Therefore, Λ is a random topological isomorphism. This completes the proof.

Corollary 2.3.8 *Every random topologically isomorphism preserves completeness.*

Theorem 2.3.9 *Every linear operator $\Lambda : (X, \mu, T) \longrightarrow (Y, \nu, T')$, where $\dim X < \infty$, but other is not necessarily finite dimensional, is continuous.*

Proof. If we define

$$\eta_x(t) = T'(\mu_x(t), \nu_{\Lambda x}(t)), \tag{2.3.4}$$

where $T' \gg T$. Then (X, η, T) is an RN-space since (RN1) and (RN2) are immediate from the definition and, for the triangle inequality (RN3),

$$\begin{aligned}
T(\eta_x(t), \eta_z(s)) &= T[T'(\mu_x(t), \nu_{\Lambda x}(t)), T'(\mu_z(s), \nu_{\Lambda z}(s))] \\
&\leq T'[T(\mu_x(t), \mu_z(s))T(\nu_{\Lambda x}(t), \nu_{\Lambda z}(s))] \\
&\leq T'(\mu_{x+z}(t + s), \nu_{\Lambda(x+z)}(t + s)) \\
&= \eta_{x+z}(t + s).
\end{aligned}$$

Now, let $x_n \xrightarrow{\mu} x$. Then, by Theorem 2.3.3, $x_n \xrightarrow{\eta} x$, but, by (2.3.3), since

$$\nu_{\Lambda x}(t) \geq \eta_x(t),$$

we have $\Lambda x_n \xrightarrow{\nu} \Lambda x$. Hence, Λ is continuous. This completes the proof.

Corollary 2.3.10 *Every linear isomorphism between finite dimensional RN-spaces is a topological isomorphism.*

Corollary 2.3.11 *Every finite dimensional RN-space (X, μ, T) is complete.*

Proof. By Corollary 2.3.10, (X, μ, T) and $(\mathbb{R}^n, \Phi, T)$ are random topologically isomorph. Since $(\mathbb{R}^n, \Phi, T)$ is complete and every random topological isomorphism preserves completeness, (X, μ, T) is complete.

Definition 2.3.12 Let (V, μ, T) be an RN-space, W be a linear manifold in V, and $Q : V \longrightarrow V/W$ be the natural mapping with $Qx = x + W$. For any $t > 0$, we define

$$\bar{\mu}(x + W, t) = \sup\{\mu_{x+y}(t) : y \in W\}.$$

Theorem 2.3.13 Let W be a closed subspace of an RN-space (V, μ, T). If $x \in V$ and $\epsilon > 0$, then there exists $x' \in V$ such that

$$x' + W = x + W, \quad E_{\lambda,\mu}(x') < E_{\lambda,\mu}^{-}(x + W) + \epsilon.$$

Proof. By the properties of sup, there always exists $y \in W$ such that

$$E_{\lambda,\mathcal{P}}(x + y) < E_{\lambda,\mu}^{-}(x + W) + \epsilon.$$

Now, it is enough to put $x' = x + y$.

Theorem 2.3.14 Let W be a closed subspace of an RN-space (V, μ, T) and $\bar{\mu}$ be given in the above definition. Then we have

(1) $\bar{\mu}$ is an RN-space on V/W;
(2) $\bar{\mu}_{Qx}(t) \geq \mu_x(t)$;
(3) If (V, μ, T) is a random Banach space, then so is $(V/W, \bar{\mu}, T)$.

Proof. (1) It is clear that $\bar{\mu}_{x+W}(t) > 0$. Let $\bar{\mu}_{x+W}(t) = 1$. By the definition, there exists a sequence $\{x_n\}$ in W such that $\mu_{x+x_n}(t) \longrightarrow 1$. Thus, $x+x_n \longrightarrow 0$ or, equivalently, $x_n \longrightarrow (-x)$ and since W is closed, $x \in W$ and $x+W = W$, the zero element of V/W. Now, we have

$$\begin{aligned}
\bar{\mu}_{(x+W)+(y+W)}(t) &= \bar{\mu}_{(x+y)+W}(t) \\
&\geq \mu_{(x+m)+(y+n)}(t) \\
&\geq T(\mu_{x+m}(t_1), \mu_{y+n}(t_2))
\end{aligned}$$

for all $m, n \in W$, $x, y \in V$, and $t_1 + t_2 = t$. Now, if we take the sup, then we have

$$\bar{\mu}_{(x+W)+(y+W)}(t) \geq T(\bar{\mu}_{x+W}(t_1), \bar{\mu}_{y+W}(t_2)).$$

Therefore, $\bar{\mu}$ is a random norm on V/W.

(2) By Definition 2.3.12, we have

$$\bar{\mu}_{Qx}(t) = \bar{\mu}_{x+W}(t) = \sup\{\mu_{x+y}(t) : y \in W\} \geq \mu_x(t).$$

Note that, by Lemma 2.2.15,

$$E_{\lambda,\bar{\mu}}(Qx) = \inf\{t > 0 : \bar{\mu}_{Qx}(t) > 1 - \lambda\}$$
$$\leq \inf\{t > 0 : \mu_x(t) > 1 - \lambda\} \tag{2.3.5}$$
$$= E_{\lambda,\mu}(x).$$

(3) Let $\{x_n + W\}$ be a Cauchy sequence in V/W. Then there exists $n_0 \in \mathbb{N}$ such that, for each $n \geq n_0$,

$$E_{\lambda,\bar{\mu}}((x_n + W) - (x_{n+1} + W)) \leq 2^{-n}.$$

Let $y_1 = 0$ and choose $y_2 \in W$ such that

$$E_{\lambda,\mu}(x_1 - (x_2 - y_2), t) \leq E_{\lambda,\bar{\mu}}((x_1 - x_2) + W) + \frac{1}{2}.$$

However, $E_{\lambda,\bar{\mu}}((x_1 - x_2) + W) \leq \frac{1}{2}$ and so $E_{\lambda,\mu}(x_1 - (x_2 - y_2)) \leq \left(\frac{1}{2}\right)^2$. Now, suppose that y_{n-1} has been chosen. Then choose $y_n \in W$ such that

$$E_{\lambda,\mu}((x_{n-1} + y_{n-1}) - (x_n + y_n)) \leq E_{\lambda,\bar{\mu}}((x_{n-1} - x_n) + W) + 2^{-n+1}.$$

Hence, we have

$$E_{\lambda,\mu}((x_{n-1} + y_{n-1}) - (x_n + y_n)) \leq 2^{-n+2}.$$

However, by Lemma 2.2.15, for each positive integer $m > n$ and $\lambda \in (0,1)$, there exists $\gamma \in (0,1)$ such that

$$E_{\lambda,\mu}((x_m + y_m) - (x_n + y_n)) \leq E_{\gamma,\mu}((x_{n+1} + y_{n+1}) - (x_n + y_n)) +$$
$$\cdots + E_{\gamma,\mu}((x_m + y_m) - (x_{m-1} + y_{m-1}))$$
$$\leq \sum_{i=n}^{m} 2^{-i}.$$

By Lemma 2.2.15, $\{x_n + y_n\}$ is a Cauchy sequence in V. Since V is complete, there exists x_0 in V such that $x_n + y_n \longrightarrow x_0$ in V.

On the other hand, we have

$$x_n + W = Q(x_n + y_n) \longrightarrow Q(x_0) = x_0 + W.$$

Therefore, every Cauchy sequence $\{x_n + W\}$ is convergent in V/W and so V/W is complete. Thus, $(V/W, \bar{\mu}, T)$ is a random Banach space. This completes the proof.

Theorem 2.3.15 *Let W be a closed subspace of an RN-space (V, μ, T). If two of the spaces V, W and V/W are complete, then so is the third one.*

Proof. If V is a random Banach space, then so are V/W and W. Hence, the fact that needs to be checked is that V is complete whenever both W and

V/W are complete. Suppose that W, V/W are random Banach spaces and $\{x_n\}$ is a Cauchy sequence in V. Since

$$E_{\lambda,\bar{\mu}}((x_n - x_m) + W) \leq E_{\lambda,\mu}(x_n - x_m)$$

for each $m, n \geq 1$, the sequence $\{x_n + W\}$ is a Cauchy sequence in V/W and so converges to $y + W$ for some $y \in W$. Thus, there exists $n_0 \geq 1$ such that, for each $n \geq n_0$,

$$E_{\lambda,\bar{\mu}}((x_n - y) + W) < 2^{-n}.$$

Now, by the last theorem, there exists a sequence $\{y_n\}$ in V such that

$$y_n + W = (x_n - y) + W, \quad E_{\lambda,\mu}(y_n) < E_{\lambda,\bar{\mu}}((x_n - y) + W) + 2^{-n}.$$

Thus, we have

$$\lim_{n \to \infty} E_{\lambda,\mu}(y_n) \leq 0$$

and so, by Lemma 2.2.15, $\mu_{y_n}(t) \to 1$ for any $t > 0$, that is, $\lim_{n \to \infty} y_n = 0$. Therefore, $\{x_n - y_n - y\}$ is a Cauchy sequence in W and so it is convergent to a point $z \in W$. This implies that $\{x_n\}$ converges to $z + y$ and hence V is complete. This completes the proof.

Theorem 2.3.16 (Open Mapping Theorem) *If T is a random bounded linear operator from an RN-space (V, μ, T) onto an RN-space (V', ν, T), then T is an open mapping.*

Proof. The theorem will be proved by the following steps:

Step 1: Let E be a neighborhood of the 0 in V. We show that $0 \in \left(\overline{T(E)}\right)^o$. Let W be a balanced neighborhood of 0 such that $W + W \subset E$. Since $T(V) = V'$ and W is absorbing, it follows that $V' = \cap_n T(nW)$ and so there exists $n_0 \geq 1$ such that $\overline{T(n_0 W)}$ has a nonempty interior. Therefore, we have

$$0 \in \left(\overline{T(W)}\right)^o - \left(\overline{T(W)}\right)^o.$$

On the other hand, we have

$$\left(\overline{T(W)}\right)^o - \left(\overline{T(W)}\right)^o \subset \overline{T(W)} - \overline{T(W)} = \overline{T(W)} + \overline{T(W)}$$
$$\subset \overline{T(E)}.$$

Thus, the set $\overline{T(E)}$ includes the neighborhood $\left(\overline{T(W)}\right)^o - \left(\overline{T(W)}\right)^o$ of 0.

Step 2: We show $0 \in (T(E))^o$. Since $0 \in E$ and E is an open set, there exist $0 < \alpha < 1$ and $t_0 \in (0, \infty)$ such that $B_0(\alpha, t_0) \subset E$. However, $0 < \alpha < 1$ and so a sequence $\{\epsilon_n\}$ can be found such that

$$T^{m-n}(1 - \epsilon_{n+1}, \cdots, 1 - \epsilon_m) \to 1$$

and

$$1 - \alpha < \lim_n T^{n-1}(1 - \epsilon_1, 1 - \epsilon_n),$$

in which $m > n$.

On the other hand, $0 \in \overline{T(B_0(\epsilon_n, t'_n))}$, where $t'_n = \frac{1}{2^n}t_0$, and so, by Step 1, there exist $0 < \sigma_n < 1$ and $t_n > 0$ such that

$$B_0(\sigma_n, t_n) \subset \overline{T(B_0(\epsilon_n, t'_n))}.$$

Since the set $\{B_0(r, 1/n)\}$ is a countable local base at zero and $t'_n \longrightarrow 0$ as $n \longrightarrow \infty$, t_n and σ_n can be chosen such that $t_n \longrightarrow 0$ and $\sigma_n \longrightarrow 0$ as $n \to \infty$.

Now, we show that

$$B_0(\sigma_1, t_1) \subset (T(E))^\circ.$$

Suppose that $y_0 \in B_0(, \sigma_1, t_1)$. Then, $y_0 \in \overline{T(B_0(\epsilon_1, t'_1))}$ and so for any $0 < \sigma_2$ and $t_2 > 0$, the ball $B_{y_0}(\sigma_2, t_2)$ intersects $T(B_0(\epsilon_1, t'_1))$. Therefore, there exists $x_1 \in B_0(\epsilon_1, t'_1)$ such that $Tx_1 \in B_{y_0}(\sigma_2, t_2)$, that is,

$$\nu_{y_0 - Tx_1}(t_2) > 1 - \sigma_2$$

or, equivalently,

$$y_0 - Tx_1 \in B_0(\sigma_2, t_2) \subset \overline{T(B_0(\epsilon_1, t'_1))}.$$

By the similar argument, there exists $x_2 \in B_0(\epsilon_2, t'_2)$ such that

$$\nu_{y_0 - (Tx_1 + Tx_2)}(t_3) = \nu_{(y_0 - Tx_1) - Tx_2}(t_3) > 1 - \sigma_3.$$

If this process is continued, it leads to a sequence $\{x_n\}$ such that

$$x_n \in B_0(\epsilon_n, t'_n), \quad \nu_{y_0 - \sum_{j=1}^{n-1} Tx_j}(t_n) > 1 - \sigma_n.$$

Now, if $n, m \geq 1$ and $m > n$, then we have

$$\mu_{\sum_{j=1}^n x_j - \sum_{j=n+1}^m x_j}(t) = \mu_{\sum_{j=n+1}^m x_j}(t)$$
$$\geq T^{m-n}(\mu_{x_{n+1}}(t_{n+1}), \mu_{x_m}(t_m)),$$

where $t_{n+1} + t_{n+2} + \cdots + t_m = t$. Put $t'_0 = \min\{t_{n+1}, t_{n+2}, \cdots, t_m\}$. Since $t'_n \longrightarrow 0$, there exists $n_0 \geq 1$ such that $0 < t'_n \leq t'_0$ for all $n > n_0$. Therefore, for all $m > n$, we have

$$T^{m-n}(\mu_{x_{n+1}}(t'_0), \mu_{x_m}(t'_0)) \geq T^{m-n}(\mu_{x_{n+1}}(t'_{n+1}), \mu_{x_m}(t'_m))$$
$$\geq T^{m-n}(1 - \epsilon_{n+1}, 1 - \epsilon_m)$$

and so

$$\lim_{n \longrightarrow \infty} \mu_{\sum_{j=n+1}^{m} x_j}(t) \geq \lim_{n \longrightarrow \infty} T^{m-n}(1 - \epsilon_{n+1}, 1 - \epsilon_m) = 1,$$

that is,

$$\mu_{\sum_{j=n+1}^{m} x_j}(t) \longrightarrow 1$$

for all $t > 0$. Thus, the sequence $\left\{ \sum_{j=1}^{n} x_j \right\}$ is a Cauchy sequence and so the series $\left\{ \sum_{j=1}^{\infty} x_j \right\}$ converges to a point $x_0 \in V$ since V is a complete space. For any fixed $t > 0$, there exists $n_0 \geq 1$ such that $t > t_n$ for all $n > n_0$ since $t_n \longrightarrow 0$. Thus, we have

$$\nu_{y_0 - T\left(\sum_{j=1}^{n-1} x_j\right)}(t) \geq \nu_{y_0 - T\left(\sum_{j=1}^{n-1} x_j\right)}(t_n)$$
$$\geq 1 - \sigma_n$$

and so

$$\nu_{y_0 - T\left(\sum_{j=1}^{n-1} x_j\right)}(t) \longrightarrow 1.$$

Therefore, we have

$$y_0 = \lim_{n \to \infty} T\left(\sum_{j=1}^{n-1} x_j \right) = T\left(\lim_{n \to \infty} \sum_{j=1}^{n-1} x_j \right) = Tx_0.$$

But, we have

$$\mu_{x_0}(t_0) = \lim_{n \to \infty} \mu_{\sum_{j=1}^{n} x_j}(t_0)$$
$$\geq T^n(\lim_{n \to \infty}(\mu_{x_1}(t_1'), \mu_{x_n}(t_n')))$$
$$\geq \lim_{n \to \infty} T^{n-1}(1 - \epsilon_1, \cdots, 1 - \epsilon_n)$$
$$> 1 - \alpha.$$

Therefore, $x_0 \in B_0(\alpha, t_0)$.

Step 3: Let G be an open subset of V and $x \in G$. Then, we have

$$T(G) = Tx + T(-x + G) \supset Tx + (T(-x + G))^\circ.$$

Hence, $T(G)$ is open since it includes a neighborhood of each of its point. This completes the proof.

Corollary 2.3.17 *Every one-to-one random bounded linear operator from a random Banach space onto a random Banach space has a random bounded converse.*

Theorem 2.3.18 (Closed Graph Theorem) *Let T be a linear operator from a random Banach space (V, μ, T) into a random Banach space (V', ν, T). Suppose that, for every sequence $\{x_n\}$ in V such that $x_n \longrightarrow x$ and $Tx_n \longrightarrow y$ for some elements $x \in V$ and $y \in V'$, it follows that $Tx = y$. Then T is random bounded.*

Proof. For any $t > 0$, $x \in X$ and $y \in V'$, define

$$\Phi_{(x,y)}(t) = T'(\mu_x(t), \nu_y(t)),$$

where $T' \gg T$.

First, we show that $(V \times V', \Phi, T)$ is a complete RN-space. The properties of (RN1) and (RN2) are immediate from the definition. For the triangle inequality (RN3), suppose that $x, z \in V$, $y, u \in V'$, and $t, s > 0$. Then, we have

$$
\begin{aligned}
T(\Phi_{(x,y)}(t), \Phi_{(z,u)}(s)) &= T[T'(\mu_x(t), \nu_y(t)), T'(\mu_z(s), \nu_u(s))] \\
&\leq T'[T(\mu_x(t), \mu_z(s)), T(\nu_y(t), \nu_u(s))] \\
&\leq T'(\mu_{x+z}(t+s), \nu_{y+u}(t+s)) \\
&= \Phi_{(x+z, y+u)}(t+s).
\end{aligned}
$$

Now, if $\{(x_n, y_n)\}$ is a Cauchy sequence in $V \times V'$, then, for any $\epsilon > 0$ and $t > 0$, there exists $n_0 \geq 1$ such that

$$\Phi_{(x_n, y_n) - (x_m, y_m)}(t) > 1 - \epsilon$$

for all $m, n > n_0$. Thus, for all $m, n > n_0$, we have

$$
\begin{aligned}
T'(\mu_{x_n - x_m}(t), \nu_{y_n - y_m}(t)) &= \Phi_{(x_n - x_m, y_n - y_m)}(t) \\
&= \Phi_{(x_n, y_n) - (x_m, y_m)}(t) \\
&> 1 - \epsilon.
\end{aligned}
$$

Therefore, $\{x_n\}$ and $\{y_n\}$ are Cauchy sequences in V and V', respectively, and there exist $x \in V$ and $y \in V'$ such that $x_n \longrightarrow x$ and $y_n \longrightarrow y$ and so $(x_n, y_n) \longrightarrow (x, y)$. Hence $(V \times V', \Phi, T)$ is a complete RN-space. The remainder of the proof is the same as the classical case. This completes the proof.

Chapter 3
Random compact operators

Random compact operators are useful to study random differentiation and random integral equations. In this chapter, we define the random norm of random bounded operators and study random norms of differentiation operators and integral operators. The definition of random norm of random bounded operators lets us to study the random operator theory.

Note that, in [6] the authors proved that every RN-space is a topological vector space (see also Theorem 2 of [32] and [48]).

Theorem 3.0.1 (Continuity and boundedness) *Let* (X, μ, T) *and* (Y, ν, T') *be RN-spaces, in which* $T, T' \in \Sigma$. *Let* $\Lambda : X \longrightarrow Y$ *be a linear operator. Then:*

(a) Λ *is continuous if and only if* Λ *is random bounded.*
(b) *If* Λ *is continuous at a single point, it is continuous.*

Proof. See Section 2 and Theorem 2.3.5.

3.1 Random norm of operators

Let (X, μ, T) and (Y, μ, T) be RN-spaces and $\Lambda : X \longrightarrow Y$ be a random bounded linear operator. Define

$$\eta(\Lambda) = \inf\{h > 0 : \quad \mu_{\Lambda x}(t) \geq \mu_{hx}(t)\}, \tag{3.1.1}$$

for each $x \in X$ and $t > 0$. $\eta(\Lambda)$ is called the operator random norm.

Lemma 3.1.1 *Let* (X, μ, T) *and* (Y, μ, T) *be RN-spaces and* $\Lambda : X \longrightarrow Y$ *be a random bounded linear operator. Then*

$$\mu_{\Lambda x}(t) \geq \mu_{\eta(\Lambda)x}(t), \tag{3.1.2}$$

for each $x \in X$ *and* $t > 0$.

http://dx.doi.org/10.1016/B978-0-12-805346-1.50003-2

Proof. Since $\Lambda : X \longrightarrow Y$ is a random bounded linear operator, then by (3.1.1) there exists a nonincreasing sequence $\{h_n\}$ converges to $\eta(\Lambda)$ and satisfies at

$$\mu_{\Lambda x}(t) \geq \mu_{h_n x}(t) \qquad (3.1.3)$$

for each $x \in X$ and $t > 0$. Take the limit on n from the last inequality, we get (3.1.2).

Example 3.1.2 Let (X, μ, T) be an RN-space. The identity operator $I : X \longrightarrow X$ is random bounded and

$$\eta(I) = \inf\{h > 0 : \quad \mu_{Ix}(t) = \mu_x(t)\} = 1$$

for each $x \in X$ and $t > 0$.

Example 3.1.3 Let (X, μ, T) and (Y, μ, T) be RN-spaces. The zero operator $0 : X \longrightarrow Y$ is random bounded and

$$\eta(0) = \inf\{h > 0 : \quad \mu_{0(x)}(t) = \mu_0(t) = 1\} = 0$$

for each $x \in X$ and $t > 0$.

Example 3.1.4 (Differentiation operator)
Let X be the RN-space of all polynomials on $J = [0, 1]$ with random norm given

$$\mu_x(t) = \begin{cases} 0, & \text{if } t \leq 0, \\ \min_{p \in J} \frac{t}{t + |x(p)|}, & \text{if } t > 0. \end{cases}$$

A differentiation operator D is defined on X by

$$Dx(p) = x'(p),$$

where the prime denotes differentiation with respect to p. This operator is linear but not random bounded. Indeed, let $x_n(p) = p^n$ where $n \in \mathbb{N}$. Then,

$$\mu_x(t) = \min_{p \in J} \frac{t}{t + |x(p)|} = \frac{t}{t + 1},$$

for $t > 0$ and

$$Dx_n(p) = np^{n-1}.$$

Then

$$\mu_{Dx}(t) = \min_{p \in J} \frac{t}{t + np^{n-1}} = \frac{t}{t + n},$$

for $t > 0$ and $n \in \mathbb{N}$. Now

$$\eta(D) = \inf\left\{h > 0 : \quad \frac{t}{t + n} \geq \frac{t}{t + h}\right\} = n.$$

Note that, n depended on the choice of $x \in X$.

Example 3.1.5 (Integral operator) Let X be the RN-space of all continuous functions on $J = [0, 1]$, that is, $C[0, 1]$ with random norm given

$$\mu_x(t) = \begin{cases} 0, & \text{if } t \leq 0, \\ \min_{p \in J} \frac{t}{t + |x(p)|}, & \text{if } t > 0. \end{cases}$$

We can define an integral operator

$$S : C[0, 1] \to C[0, 1]$$

by $y = Sx$ where

$$y(p) = \int_0^1 \kappa(p, \alpha) x(\alpha) d\alpha.$$

Here κ is a given function, which is called the kernel of S and is assumed to be continuous on the closed square $G = J \times J$ in the $p\alpha$-plane, where $J = [0, 1]$. This operator is linear and random bounded. The continuity of κ on the closed square implies that κ is bounded, say, $\kappa(p, \alpha) \leq k$ for all $(p, \alpha) \in G$, where k is a positive real number. Then,

$$\mu_x(t) = \min_{p \in J} \frac{t}{t + |x(p)|} = \frac{t}{t + 1},$$

for $t > 0$ and

$$\mu_{Sx}(t) = \min_{p \in J} \frac{t}{t + |\int_0^1 \kappa(p, \alpha) x(\alpha) d\alpha|}$$

$$\geq \min_{p \in J} \frac{t}{t + \int_0^1 |k||x(\alpha)| d\alpha}$$

$$\geq \min_{p \in J} \frac{t}{t + |k||x(p)|}$$

$$\geq \mu_{kx}(t)$$

for $t > 0$, that is, the integral operator S is random bounded.

Theorem 3.1.6 *Let (X, μ, T) be an RN-space, in which $T \in \Sigma$ and X is finite dimensional on the field $(\mathbb{F}, \mu', T)$, then every linear operator on X is random bounded.*

Proof. Let $\dim X = n$ and $\{e_1, ..., e_n\}$ a basis for X. We take any

$$x = \sum_{j=1}^n \alpha_j e_j,$$

and consider any linear operator Λ on X. Since Λ is linear,

$$\mu_{\Lambda x}(t) = \mu_{\sum_{j=1}^n \alpha_j \Lambda e_j}(t)$$

for $t > 0$. By Theorem 6.1 of [2] and since $T \in \Sigma$, for every $\lambda \in (0,1)$, there exist $\gamma \in (0,1)$ and $K_0 \in \mathbb{F}$ such that

$$E_{\lambda,\mu'}(K_0) \geq 1$$

and

$$E_{\lambda,\mu}(\Lambda x) = E_{\lambda,\mu}(\sum_{j=1}^{n} \alpha_j \Lambda e_j)$$

$$\leq \sum_{j=1}^{n} E_{\gamma,\mu}(\alpha_j \Lambda e_j)$$

$$\leq \sum_{j=1}^{n} |\alpha_j| \max_{1 \leq j \leq n} E_{\gamma,\mu}(\Lambda e_j)$$

$$\leq \sum_{j=1}^{n} |\alpha_j| M_0 E_{\lambda,\mu'}(K_0)$$

$$\leq E_{\lambda,\mu'}(M_0 K_0 \sum_{j=1}^{n} |\alpha_j|)$$

$$\leq E_{\lambda,\mu}(M_0 K_0 c x)$$

in which $M_0 = \max_{1 \leq j \leq n} E_{\gamma,\mu}(\Lambda e_j)$. Put $M_0 K_0 c = h$, by Theorem 3.0.1, Λ is random bounded.

Corollary 3.1.7 (Continuity, null space) *Let (X, μ, T) and (Y, μ, T) be RN-spaces. Let $\Lambda : X \longrightarrow Y$ be an random bounded linear operator. Then:*

(a) *$x_n \to x$ implies $\Lambda x_n \to \Lambda x$.*
(b) *The null space $\mathcal{N}(\Lambda) = \{x \in X : \Lambda x = 0\}$ is closed.*

Proof.

(a) Since $\Lambda : X \longrightarrow Y$ is a random bounded linear operator, we have

$$\mu_{\Lambda x_n - \Lambda x}(t) = \mu_{\Lambda(x_n - x)}(t)$$
$$\geq \mu_{\eta(\Lambda)(x_n - x)}(t)$$
$$\to 1,$$

for every $t > 0$.

(b) Let $x \in \overline{\mathcal{N}(\Lambda)}$, then there exists a sequence $\{x_n\}$ in $\mathcal{N}(\Lambda)$ such that $x_n \to x$. By part (a) of this corollary, we have $\Lambda x_n \to \Lambda x$. Since $\Lambda x_n = 0$, then $\Lambda x = 0$ which implies that $x \in \mathcal{N}(\Lambda)$. Since $x \in \overline{\mathcal{N}(\Lambda)}$ was arbitrary, $\mathcal{N}(\Lambda)$ is closed.

3.2 Random Operator Space

Let (X, μ, T) and (Y, μ, T) be RN-spaces. In this section, first, we consider the set $B(X, Y)$ consisting of all random bounded linear operators from X into Y. We want to show that $B(X, Y)$ can itself be made into a normed space. The whole matter is quite simple. First of all, $B(X, Y)$ becomes a vector space if we define the sum $\Lambda_1 + \Lambda_2$ of two operators $\Lambda_1, \Lambda_2 \in B(X, Y)$ in a natural way by

$$(\Lambda_1 + \Lambda_2)x = \Lambda_1 x + \Lambda_2 x$$

and the product $\alpha \Lambda$ of $\Lambda \in B(X, Y)$ and a scalar α by

$$(\alpha \Lambda)x = \alpha \Lambda x.$$

Note that, if (3.1.1) holds, then for every $\lambda \in (0, 1)$ we have

$$\eta(\Lambda) = \inf\{h > 0 : \; E_{\lambda,\mu}(\Lambda x) \leq E_{\lambda,\mu}(hx)\} \tag{3.2.1}$$

and therefore

$$E_{\lambda,\mu}(\Lambda x) \leq \eta(\Lambda) E_{\lambda,\mu}(x) \tag{3.2.2}$$

for $x \in X$. Then

$$E_\mu(\Lambda x) \leq \eta(\Lambda) E_\mu(x) \tag{3.2.3}$$

for $x \in X$ in which

$$E_\mu(\Lambda x) = \sup_{\lambda \in (0,1)} E_{\lambda,\mu}(\Lambda x) < \infty. \tag{3.2.4}$$

Theorem 3.2.1 *Let (X, μ, T) and (Y, μ, T) be RN-spaces, in which $T \in \Sigma$. The vector space $B(X, Y)$ of all random bounded linear operators from X into Y is itself a normed space with norm defined by (3.1.1) whenever $E_\mu(x) < \infty$.*

Proof. In Example, 3.1.3 we showed that $\eta(0) = 0$. Now, if $\eta(\Lambda) = 0$ we have $\mu_{\Lambda x}(t) = 1$ for each $x \in X$ and $t > 0$, which implies that $\Lambda x = 0$ and $\Lambda = 0$. On the other hand,

$$\begin{aligned}
\eta(\alpha \Lambda) &= \inf\{h > 0 : \; \mu_{\alpha \Lambda x}(t) \geq \mu_{hx}(t)\} \\
&= \inf\{h > 0 : \; \mu_{\Lambda x}(t) \geq \mu_{\frac{h}{\alpha}x}(t)\} \\
&= |\alpha| \inf\{h > 0 : \; \mu_{\Lambda x}(t) \geq \mu_{hx}(t)\} \\
&= |\alpha| \eta(\Lambda).
\end{aligned}$$

Now, we prove triangle inequality for η. Let $\Lambda, \Gamma \in B(X, Y)$, then

$$\mu_{(\Lambda + \Gamma)x}(t) \geq \mu_{\eta(\Lambda + \Gamma)x}(t),$$

for each $x \in X$ and $t > 0$. For every $\lambda \in (0,1)$ there exists $\gamma \in (0,1)$ such that both

$$E_{\lambda,\mu}((\Lambda + \Gamma)x) \leq \eta(\Lambda + \Gamma)E_{\lambda,\mu}(x)$$

which implies that,

$$E_\mu((\Lambda + \Gamma)x) \leq \eta(\Lambda + \Gamma)E_\mu(x) \tag{3.2.5}$$

and

$$E_{\lambda,\mu}((\Lambda + \Gamma)x) \leq E_{\gamma,\mu}(\Lambda x) + E_{\gamma,\mu}(\Gamma x)$$
$$\leq [\eta(\Lambda) + \eta(\Gamma)]E_{\gamma,\mu}(x)$$

which implies that,

$$E_\mu((\Lambda + \Gamma)x) \leq [\eta(\Lambda) + \eta(\Gamma)]E_\mu(x) \tag{3.2.6}$$

for each $x \in X$. From (3.2.5) and (3.2.6) we have

$$\eta(\Lambda + \Gamma) \leq \eta(\Lambda) + \eta(\Gamma).$$

Theorem 3.2.2 *Let (X, μ, T) and (Y, μ, T) be RN-spaces, in which $T \in \Sigma$ and X. If Y is a complete RN-space then $(B(X,Y), \eta)$ is complete whenever $E_\mu(x) < \infty$.*

Proof. We consider an arbitrary Cauchy sequence $\{\Lambda_n\}$ in $(B(X,Y), \eta)$ and show that $\{\Lambda_n\}$ converges to an operator $\Lambda \in B(X,Y)$. Since $\{\Lambda_n\}$ is Cauchy, for every $h > 0$, there exists $N \in \mathbb{N}$ such that if $m, n \geq N$ then

$$\eta(\Lambda_n - \Lambda_m) < h,$$

or

$$\eta(\Lambda_n - \Lambda_m) \to 0$$

whenever m, n tend to ∞. For all $x \in X$ and $t > 0$ we have

$$\mu_{\Lambda_n x - \Lambda_m x}(t) = \mu_{(\Lambda_n - \Lambda_m)x}(t)$$
$$\geq \mu_x \left(\frac{t}{\eta(\Lambda_n - \Lambda_m)} \right) \tag{3.2.7}$$
$$\to 1$$

whenever m, n tend to ∞. Then, the sequence $\{\Lambda_n x\}$ is Cauchy in complete RN-space (Y, μ, T) and so converges to $y \in Y$ depends on the choice of $x \in X$. This defines an operator $\Lambda : X \to Y$, where $y = \Lambda x$. The operator Λ is linear since

$$\lim \Lambda_n(\alpha x + \beta z) = \lim \alpha \Lambda_n x + \lim \beta \Lambda_n z = \alpha \lim \Lambda_n x + \beta \lim \Lambda_n z$$

for $x, z \in X$ and scalars α, β.

Now, we show that Λ is random bounded and $\Lambda_n \to \Lambda$. For every $m, n \geq N$ we have

$$
\begin{aligned}
\mu_{\Lambda_n x - \Lambda_m x}(t) &= \mu_{(\Lambda_n - \Lambda_m)x}(t) \\
&\geq \mu_x\left(\frac{t}{\eta(\Lambda_n - \Lambda_m)}\right) \quad\quad (3.2.8) \\
&\geq \mu_x\left(\frac{t}{h}\right).
\end{aligned}
$$

On the other hand, $\Lambda_m x \to \Lambda x$ when m tends to ∞. Using the continuity of the random norm, we obtain from (3.2.7), for every $n > N$, $x \in X$, and $t > 0$

$$
\begin{aligned}
\mu_{\Lambda_n x - \Lambda x}(t) &= \lim_{m \to \infty} \mu_{(\Lambda_n - \Lambda_m)x}(t) \\
&\geq \lim_{m \to \infty} \mu_x\left(\frac{t}{\eta(\Lambda_n - \Lambda_m)}\right) \quad\quad (3.2.9) \\
&\geq \mu_x\left(\frac{t}{h}\right).
\end{aligned}
$$

This shows that $(\Lambda_n - \Lambda)$ with $n > N$ is a random bounded linear operator. Since Λ_n is random bounded, $\Lambda = \Lambda_n - (\Lambda_n - \Lambda)$ is random bounded, that is, $\Lambda \in B(X, Y)$. From (3.2.9) we have

$$
\mu_x\left(\frac{t}{\eta(\Lambda_n - \Lambda)}\right) \geq \mu_x\left(\frac{t}{h}\right).
$$

Then

$$
\eta(\Lambda_n - \Lambda) \leq h
$$

for every $n > N$. Hence

$$
\Lambda_n \xrightarrow{\eta} \Lambda.
$$

A functional is an operator whose range lies on the real line $\mathbb{R}$ or in the complex plane $\mathbb{C}$. A random bounded linear functional is a random bounded linear operator with range in the scalar field of the RN-space (X, μ, T). It is of basic importance that the set of all linear functionals defined on a vector space X can itself be made into a vector space. Let $(\mathbb{F}, \mu', T)$ be an RN-space ($\mathbb{F} = \mathbb{R}$ or $\mathbb{C}$). The set $X' = B(X, \mathbb{F})$ is said to be random dual space. The random dual space X' is Banach space with the norm η.

3.3 Compact Operators

Definition 3.3.1 (R-Compact linear operator) *Let (X, μ, T) and (Y, μ, T') be RN-spaces. An operator $\Lambda : X \longrightarrow Y$ is called an R-compact linear operator if Λ is linear and if for every random bounded subset M of X, the closure $\overline{\Lambda(M)}$ is R-compact.*

Lemma 3.3.2 *Let (X, μ, T) and (Y, μ, T) be RN-spaces. Then, every R-compact linear operator $\Lambda : X \longrightarrow Y$ is random bounded, hence continuous.*

Proof. Let U be a random bounded set, then there exist $r_0 \in (0, 1)$ and $t_0 > 0$ such that

$$\mu_x(t_0) \geq 1 - r_0,$$

for every $x \in U$. On the other hand, $\overline{\Lambda(U)}$ is R-compact and by Theorem 2.2.6 is R-bounded, then there exist $r_1 \in (0, 1)$ and $t_1 > 0$ such that

$$\mu_{\Lambda x}(t_1) \geq 1 - r_1,$$

for every $x \in U$. By the intermediate value theorem, there exists a positive real number h_0 such that

$$\mu_{\Lambda x}(h_0 t_0) \geq \mu_x(t_0),$$

for every $x \in U$ (note that by the last inequality h_0 cannot tend to zero), and so $\eta(\Lambda) < \infty$. Hence, Λ is random bounded and by Theorem 3.0.1 is continuous.

Chapter 4
Random Banach algebras

In this chapter, at first, we consider the concept of random Banach algebras and random compact operators. Then, we apply a fixed point theorem to solve the operator equation $AxBx = x$ in the random Banach algebras under a nonlinear contraction.

4.1 Random normed algebra

Definition 4.1.1 *A random normed algebra* (X, μ, T, T') *is a random normed space* (X, μ, T) *with algebraic structure such that*

(RN-4) $\mu_{xy}(ts) \geq T'(\mu_x(t), \mu_y(s))$ *for all* $x, y \in X$ *and all* $t, s > 0$. *In which* T' *is a continuous t-norm.*

Every normed algebra $(X, \|\cdot\|)$ defines a random normed algebra (X, μ, T_M, T_P), where

$$\mu_x(t) = \frac{t}{t + \|x\|}$$

for all $t > 0$, if and only if

$$\|xy\| \leq \|x\|\|y\| + s\|y\| + t\|x\| \qquad (x, y \in X; \quad t, s > 0).$$

This space is called the induced random normed algebra.

Note that the last example is held with (X, μ, T_P, T_P).

Theorem 4.1.2 (The Schauder fixed point theorem) *Let* K *be a convex subset of a topological vector space* X *and* A *is a continuous mapping of* K *into itself so that* $A(K)$ *is contained in an R-compact subset of* K, *then* A *has a fixed point.*

Proof. The proof is depended to topological vector space properties; therefore, we omit it.

Random Operator Theory
http://dx.doi.org/10.1016/B978-0-12-805346-1.50004-4
Copyright © 2016 Elsevier Ltd. All rights reserved.

Definition 4.1.3 Let (X, μ, T) be an RN-space and $A \subset X$. We say A is *totally bounded* if for each $0 < r < 1$ and $t > 0$ there exists a finite subset S of X such that

$$A \subseteq \bigcup_{x \in S} B_x(r, t).$$

Lemma 4.1.4 *Let (X, μ, T) be an RN-space and $A \subset X$. Then*

(a) *If $\overline{A}$ is R-compact, A is totally bounded.*
(b) *If A is totally bounded and X is complete, $\overline{A}$ is R-compact.*

Proof. (a) We assume that $\overline{A}$ is compact and show that, any fixed $0 < r_0 < 1$ and $t_0 > 0$ being given, there exists a finite subset S of X such that

$$A \subseteq \bigcup_{x \in S} B_x(r_0, t).$$

If $A = \emptyset$, then $S = \emptyset$. If $A \neq \emptyset$, we pick any $x_1 \in A$. If $\mu_{x_1 - z}(t_0) > 1 - r_0$ for all $z \in A$, then $\{x_1\} = S$. Otherwise, let $x_2 \in A$ be such that $\mu_{x_1 - x_2}(t_0) \leq 1 - r_0$. If for all $z \in A$,

$$\mu_{x_j - z}(t_0) > 1 - r_0 \quad \text{for} \quad j = 1 \text{ or } 2. \tag{4.1.1}$$

Then $\{x_1, x_2\} = S$. Otherwise let $z = x_3 \in A$ be a point not satisfying (4.1.1). If for all $z \in A$,

$$\mu_{x_j - z}(t_0) > 1 - r_0 \quad \text{for} \quad j = 1, 2 \text{ or } 3,$$

then $\{x_1, x_2, x_3\} = S$. Otherwise we continue by selecting an $x_4 \in A$, etc. We assert the existence of a positive integer n such that the set $\{x_1, x_2, ..., x_n\} = S$ obtained after n such steps. In fact, if there were no such n, our construction would yield a sequence $\{x_j\}$ satisfying

$$\mu_{x_j - x_k}(t_0) \leq 1 - r_0 \quad \text{for} \quad j \neq k.$$

Obviously, $\{x_j\}$ could not have a subsequence which is Cauchy. Hence, $\{x_j\}$ could not have a subsequence which converges in X. But this contradicts the compactness of $\overline{A}$ because $\{x_j\}$ lies in A, by the construction. Hence, there must be a finite set S for A. Since $0 < r_0 < 1$ and t_0 were arbitrary, we conclude that A is totally bounded.

(b) See [17].

Further, the operator Λ is called *completely continuous* if it is continuous and totally bounded.

4.2 Fixed point theorem

In this section, we assume that (X, μ, T) and (Y, μ, T) are Banach normed spaces. Let Φ be the set of all nondecreasing functions

$$\phi : [0, \infty) \longrightarrow [0, \infty).$$

Here, $\phi^n(t)$ denotes the nth iterative function of $\phi(t)$. Our results extend and improve some results of [3, 13].

Lemma 4.2.1 ([18, 19]) *If $\phi \in \Phi$ and satisfies $\sum_{j=1}^{\infty} \phi^j(t) < \infty$ for $t > 0$ then $\phi(t) < t$ for $t > 0$.*

A mapping $A : X \longrightarrow X$ is called $\mathcal{D}$-Lipschitzian if

$$\mu_{A(x)-A(y)}(\phi(t)) \geq \mu_{x-y}(t) \qquad (4.2.1)$$

for $x, y \in X$ and $t > 0$, where $\phi \in \Phi$.
  We call the function ϕ a $\mathcal{D}$-function of A on X.

Lemma 4.2.2 *Let (X, μ, T) be an RN-space, let the random norm $\mu_x(.)$ be continues for $t > 0$ and $0 < a < 1$. Then there exists $k \geq 1$ such that*

$$T(\mu_x(t), a) = \mu_x\left(\frac{t}{k}\right),$$

for $x \in X$.

Proof. Let $x \in X$, $t > 0$, and $0 < a < 1$ be fixed. Then $T(\mu_x(t), a) \in (0, 1)$. By the intermediate value theorem, there exists an $\ell > 0$ such that $T(\mu_x(t), a) = \mu_x(\ell)$. Put $\ell = \frac{t}{k}$ in which $k = k(a)$.

Theorem 4.2.3 *Let (X, μ, T) be a complete RN-space and let the random norm $\mu_x(.)$ be continuous for $t > 0$, in which $T \in \Sigma$, and let $A : X \longrightarrow X$ such that*

$$\mu_{Ax_1-Ax_2}(\phi(t)) \geq T(\mu_{x_1-x_2}(t), a),$$

for all $x_1, x_2 \in X$, $t > 0$, $\mathcal{D}$-function ϕ and some $a \in (0, 1)$ such that $\sum_{j=1}^{\infty} k^j(a)\phi^j(t) < \infty$. Then A has a unique fixed point.

Proof. Choose $x_0 \in X$ and let $t > 0$. Choose $x_1 \in X$ with $Ax_0 = x_1$. In general, choose x_{n+1} such that $Ax_n = x_{n+1}$. Now,

$$\mu_{Ax_n - Ax_{n+1}}(\phi^{n+1}(t)) \geq T(\mu_{x_n - x_{n+1}}(\phi^n(t)), a)$$
$$= T(\mu_{Ax_{n-1} - Ax_n}(\phi^n(t)), a)$$
$$= \mu_{Ax_{n-1} - Ax_n}\left(\frac{\phi^n(t)}{k}\right)$$
$$\geq$$
$$\vdots$$
$$\geq \mu_{x_0 - x_1}\left(\frac{t}{k^{n+1}}\right).$$

Note for each $\lambda \in (0, 1)$ that (see Lemma 1.9. of [41])

$$E_{\lambda,\mu}(Ax_n - Ax_{n+1}) = \inf\{\phi^{n+1}(t) > 0 : \mu_{Ax_n - Ax_{n+1}}(\phi^{n+1}(t)) > 1 - \lambda\}$$
$$\leq \inf\{\phi^{n+1}(t) > 0 : \mu_{x_0 - x_1}\left(\frac{t}{k^{n+1}}\right) > 1 - \lambda\}$$
$$\leq \phi^{n+1}\left(\inf\left\{t > 0 : \mu_{x_0 - x_1}\left(\frac{t}{k^{n+1}}\right) > 1 - \lambda\right\}\right)$$
$$= k^{n+1}\phi^{n+1}(E_{\lambda,\mu}(x_0 - x_1)),$$

in which $k \geq 1$. Now, by the proof of Lemma 1.9 of [41], the sequence $\{Ax_n\}$ is Cauchy and so is convergent since X is complete. By Theorem 2.3 of [41], A has a unique fixed point.

Theorem 4.2.4 *Let S be a closed, convex, and R-bounded subset of a Banach algebra (X, μ, T, T) in which random norm is continuous for every $t > 0$, $T \in \Sigma$ and let $A : X \longrightarrow X$, $B : S \longrightarrow X$ be two operators such that*

(a) *A is $\mathcal{D}$-Lipschitzian with a $\mathcal{D}$-function ϕ.*
(b) *B is completely continuous.*
(c) *$x = AxBy \Longrightarrow x \in S$, for all $y \in S$.*

Then the operator equation

$$AxBx = x \qquad\qquad (4.2.2)$$

has a solution, whenever $\sum_{j=1}^{\infty} k^j(a)\phi^j(\ell) < \infty$, for $\ell > 0$ in which $a = T(\mu_{B(S)}(1), \mu_{ANy}(1))$, $N : S \to X$ is a mapping defined by

$$Ny = z,$$

where $z \in X$ is the unique solution of the equation

$$z = AzBy, \quad y \in S.$$

Also $\mu_{B(S)}(1) = \inf_{y \in S} \mu_{By}(1)$.

Proof. Let $y \in S$. Now, define a mapping $A_y : X \to X$ by

$$A_y(x) = AxBy, \quad x \in X.$$

Since, for $x_i \in X$ $(i = 1, 2)$ and $t > 0$,

$$
\begin{aligned}
\mu_{A_y x_1 - A_y x_2}(\phi(t)) &= \mu_{Ax_1 B_y - Ax_2 By}(\phi(t)) \\
&\geq T(\mu_{Ax_1 - Ax_2}(\phi(t)), \mu_{By}(1)) \\
&\geq T(\mu_{x_1 - x_2}(t), \mu_{B(S)}(1)).
\end{aligned}
$$

Then, A_y is a nonlinear contraction on X with a $\mathcal{D}$-function ϕ and $\mu_{B(S)}(1) \in (0, 1)$. Now an application of a fixed point theorem 4.2.3 yields that there is a unique point $x^* \in X$ such that

$$A_y(x^*) = x$$

or, equivalently,

$$x^* = Ax^* By.$$

Since hypothesis (c) holds, we have that $x^* \in S$. We show that N is continuous. Let $\{y_n\}$ be a sequence in S converging to a point y. Since S is closed, $y \in S$. Now,

$$
\begin{aligned}
\mu_{Ny_n - Ny}(\phi(t) + \epsilon) &= \mu_{ANy_n By_n - ANyBy}(\phi(t) + \epsilon) \\
&\geq T(\mu_{ANy_n By_n - ANyBy_n}(\phi(t)), \mu_{ANyBy_n - ANyBy}(\epsilon)) \\
&\geq T(\mu_{ANy_n - ANy}(\phi(t)), \mu_{By_n}(1), \mu_{ANy}(1), \mu_{By_n - By}(\epsilon)) \\
&\geq T(\mu_{Ny_n - Ny}(t), \mu_{By_n}(1), \mu_{ANy}(1), \mu_{By_n - By}(\epsilon)),
\end{aligned}
$$

for $t > 0$ and $\epsilon \in (0, 1)$. For great $n \in \mathbb{N}$ and by continuity of B, we have

$$\mu_{Ny_n - Ny}(\phi(t) + \epsilon) \geq T(\mu_{Ny_n - Ny}(t), \mu_{By_n}(1), \mu_{ANy}(1)).$$

Take infimum on $\epsilon \in (0, 1)$, we get

$$
\begin{aligned}
\mu_{Ny_n - Ny}(\phi(t)) &\geq T(\mu_{Ny_n - Ny}(t), \mu_{B(S)}(1), \mu_{ANy}(1)) \\
&\geq T(\mu_{Ny_n - Ny}(t), a),
\end{aligned}
$$

in which $a = T(\mu_{B(S)}(1), \mu_{ANy}(1)) \in (0, 1)$. Now, by proof of Theorem 4.2.3, $\mu_{Ny_n - Ny}(\phi(t))$ tends to 1 for each $t > 0$, whenever $n \to \infty$ and consequently N is continuous on S. Next, we show that N is a compact operator on S. Let $\eta > 0$ and $t > 0$ be given. For any $z \in S$, we have

$$
\begin{aligned}
\mu_{Az}\left(2\phi\left(\frac{\sqrt{t}}{3}\right)\right) &\geq T\left(\mu_{Aa}\left(\phi\left(\frac{\sqrt{t}}{3}\right)\right), \mu_{Az - Aa}\left(\phi\left(\frac{\sqrt{t}}{3}\right)\right)\right) \\
&\geq T\left(\mu_{Aa}\left(\phi\left(\frac{\sqrt{t}}{3}\right)\right), \mu_{z - a}\left(\frac{\sqrt{t}}{3}\right)\right) \\
&\geq c,
\end{aligned}
$$

where $c = T\left(\mu_{Aa}\left(\phi\left(\frac{\sqrt{t}}{3}\right)\right), \mu_S\left(\frac{\sqrt{t}}{3}\right)\right)$ for some fixed $a \in S$. Then, we can find $r = r_{\eta,t}$ such that

$$T^3\left(\mu_S\left(\frac{t}{3}\right), \mu_{B(S)}(1), c, 1 - r\right) > 1 - \eta.$$

Since B is completely continuous, $B(S)$ is totally bounded. Then, there is a set $Y = \{y_1, ..., y_n\}$ in S such that

$$B(S) \subset \bigcup_{j=1}^{n} B_{w_j}(r, \sqrt{t}),$$

where $w_j = B(y_j)$. Therefore, for any $y \in S$, we have a $y_k \in Y$ such that

$$\mu_{By - By_k}(\sqrt{t}) > 1 - r.$$

Also we have

$$\mu_{Ny_n - Ny}(t) \geq \mu_{Ny_n - Ny}\left(\phi\left(\frac{t}{3}\right) + 2\sqrt{t}\phi\left(\frac{\sqrt{t}}{3}\right)\right)$$

$$\geq \mu_{AzBy - Az_kBy_k}\left(\phi\left(\frac{t}{3}\right) + 2\sqrt{t}\phi\left(\frac{\sqrt{t}}{3}\right)\right)$$

$$\geq T\left(\mu_{AzBy - Az_kBy}\left(\phi\left(\frac{t}{3}\right)\right), \mu_{Az_kBy - Az_kBy_k}\left(2\sqrt{t}\phi\left(\frac{\sqrt{t}}{3}\right)\right)\right)$$

$$\geq T^3\left(\mu_{Az - Az_k}\left(\phi\left(\frac{t}{3}\right)\right), \mu_{By}(1), \mu_{Az_k}\left(2\phi\left(\frac{\sqrt{t}}{3}\right)\right), \mu_{By_n - By}(\sqrt{t})\right)$$

$$\geq T^3\left(\mu_{z - z_k}\left(\frac{t}{3}\right), \mu_{B(S)}(1), c, 1 - r\right)$$

$$\geq T^3\left(\mu_S\left(\frac{t}{3}\right), \mu_{B(S)}(1), c, 1 - r\right)$$

$$\geq 1 - \eta.$$

This is true for every $y \in S$ and hence

$$N(S) \subset B_{z_i}(\eta, t),$$

where $z_i = N(y_i)$. As a result $N(S)$ is totally bounded. Since N is continuous, it is a compact operator on S. Now an application of Schauder's fixed point yields that N has a fixed point in S. Then, by the definition of N

$$x = Nx = A(Nx)Bx = AxBx,$$

and so the operator equation $x = AxBx$ has a solution in S.

Chapter 5
Fixed point theorems in random normed spaces

In this chapter, we consider some fixed point theorems in random normed spaces. In the first section, we prove the existence of tripled fixed point and tripled coincidence point theorems in RN-spaces. Next, we study the existence of a unique solution to an initial value problem, as an application to the our tripled fixed point theorem.

5.1 Tripled coincidence point theorems in RN-spaces

The aim of this section is to prove the existence of tripled fixed point and tripled coincidence point theorems in RN-spaces. Our results generalize and extend recent coupled fixed point theorems in RN-spaces.

Example 5.1.1 ([1]) Let $(X, \|\cdot\|)$ be an ordinary normed space and ϕ be an increasing and continuous function from $\mathbb{R}^+$ into $(0, 1)$ such that $\lim\limits_{t \to \infty} \phi(t) = 1$. Four typical examples of these functions are as follows:

$$\phi(t) = \frac{t}{t+1}, \phi(t) = \sin(\frac{\pi t}{2t+1}), \phi(t) = 1 - e^{-t}, \phi(t) = e^{\frac{-1}{t}}.$$

Let $T = \min$. For any $t \in (0, \infty)$, we define

$$\mu_x(t) = [\phi(t)]^{\|x\|}, \quad \forall x \in X,$$

then $(X, \mu, \min)$ is an RN-spaces.

Lemma 5.1.2 ([18]) Let $(X, \mu, \min)$ be an RN-space. Let $\{x_n\}$ be a sequence in X. If

$$\mu_{x_{n+1}-x_n}(kt) \geq \mu_{x_n-x_{n-1}}(t)$$

for some $k > 1$, $n \in \mathbb{N}$, and $t > 0$. Then, the sequence $\{x_n\}$ is Cauchy.

Random Operator Theory
http://dx.doi.org/10.1016/B978-0-12-805346-1.50005-6

Lemma 5.1.3 *Let $(X, \mu, \min)$ be an RN-space. Define*

$$Q_{x,y,z}(t) = \min(\mu_x(t), \mu_y(t), \mu_z(t))$$

for all $x, y, z \in X$ and $t > 0$. Then Q defines a random norm on $X^3 \times (0, \infty)$.

Proof. Let $Q_{x,y,z}(t) = 1$ then $\min(\mu_x(t), \mu_y(t), \mu_z(t)) = 1$, which implies that $x = y = z = 0$, the converse is trivial.

$$
\begin{aligned}
Q_{\alpha x, \alpha y, \alpha z}(t) &= \min(\mu_{\alpha x}(t), \mu_{\alpha y}(t), \mu_{\alpha z}(t)) \\
&= \min\left(\mu_x\left(\frac{t}{\alpha}\right), \mu_y\left(\frac{t}{\alpha}\right), \mu_z\left(\frac{t}{\alpha}\right)\right) \\
&= Q_{x,y,z}\left(\frac{t}{\alpha}\right)
\end{aligned}
$$

for $x, y, z \in X$, $\alpha \neq 0$ and $t > 0$.

$$
\begin{aligned}
Q_{x+x',y+y',z+z'}(t+s) &= \min(\mu_{x+x'}(t+s), \mu_{y+y'}(t+s), \mu_{z+z'}(t+s)) \\
&\geq \min(\mu_x(t), \mu_{x'}(s), \mu_y(t), \mu_{y'}(s), \mu_z(t), \mu_{z'}(s)) \\
&= \min([\mu_x(t), \mu_y(t), \mu_z(t)], [\mu_{x'}(s), \mu_{y'}(s), \mu_{z'}(s)]) \\
&= \min(Q_{x,y,z}(t), Q_{x',y',z'}(s))
\end{aligned}
$$

for $x, y, z, x', y', z' \in X$, and $t, s > 0$.

Lemma 5.1.4 *Let Q be a random norm on $X^3 \times (0, \infty)$. If*

$$Q_{x_{n+1}-x_n, y_{n+1}-y_n, z_{n+1}-z_n}(kt) \geq Q_{x_n-x_{n-1}, y_n-y_{n-1}, z_n-z_{n-1}}(t)$$

for some $k > 1$, $n \in \mathbb{N}$, and $t > 0$. Then, the sequences $\{x_n\}$, $\{y_n\}$, and $\{z_n\}$ are Cauchy.

Proof. By Lemmas 5.1.2 and 5.1.3, the proof is easy.

Definition 5.1.5 ([9]) Let X be a nonempty set. An element $(x, y, z) \in X \times X \times X$ is called a *tripled fixed point* of $F : X \times X \times X \longrightarrow X$ if

$$x = F(x, y, z), y = F(y, x, y) \text{ and } z = F(z, y, x).$$

Definition 5.1.6 Let X be a nonempty set. An element $(x, y, z) \in X \times X \times X$ is called a *tripled coincidence point* of mappings $F : X \times X \times X \longrightarrow X$ and $g : X \longrightarrow X$ if

$$g(x) = F(x, y, z), \ g(y) = F(y, x, y), \text{ and } g(z) = F(z, y, x).$$

Definition 5.1.7 ([9]) Let $(X, \preceq)$ be a partially ordered set. A mapping $F : X \times X \times X \longrightarrow X$ is said to have the *mixed monotone property* if F

is monotone nondecreasing in its first and third argument and is monotone nonincreasing in its second argument, that is, for any $x, y, z \in X$

$$x_1, x_2 \in X, x_1 \preceq x_2 \Longrightarrow F(x_1, y, z) \preceq F(x_2, y, z)$$

$$y_1, y_2 \in X, y_1 \preceq y_2 \Longrightarrow F(x, y_2, z) \preceq F(x, y_1, z)$$

and

$$z_1, z_2 \in X, z_1 \preceq z_2 \Longrightarrow F(x, y, z_1) \preceq F(x, y, z_2).$$

Definition 5.1.8 Let $(X, \preceq)$ be a partially ordered set, and $g : X \longrightarrow X$. A mapping $F : X \times X \times X \longrightarrow X$ is said to have the *mixed g-monotone property* if F is monotone g-nondecreasing in its first and third argument and is monotone g-nonincreasing in its second argument, that is, for any $x, y, z \in X$

$$x_1, x_2 \in X, \ g(x_1) \preceq g(x_2) \Longrightarrow F(x_1, y, z) \preceq F(x_2, y, z)$$

$$y_1, y_2 \in X, \ g(y_1) \preceq g(y_2) \Longrightarrow F(x, y_2, z) \preceq F(x, y_1, z)$$

and

$$z_1, z_2 \in X, \ g(z_1) \preceq g(z_2) \Longrightarrow F(x, y, z_1) \preceq F(x, y, z_2).$$

Lemma 5.1.9 ([22]) *Let X be a nonempty set and $g : X \longrightarrow X$ be a mapping. Then there exists a subset $E \subseteq X$ such that $g(E) = g(X)$ and $g : E \longrightarrow X$ is one-to-one.*

Theorem 5.1.10 *Let $(X, \mu, \min)$ be a complete RN-space, $\preceq$ be a partial order on X. Suppose that $F : X \times X \times X \longrightarrow X$ has mixed monotone property and*

$$\mu_{F(x,y,z)-F(u,v,w)}(kt) \qquad \qquad (5.1.1)$$
$$\geq \min(\mu_{x-u}(t), \mu_{y-v}(t), \mu_{z-w}(t))$$

for all those x, y, z, u, v, w in X for which $x \preceq u$, $y \succeq v$, $z \preceq w$, where $0 < k < 1$. If either

(a)   *F is continuous or*
(b)   *X has the following property:*

(bi)   *if $\{x_n\}$ is a nondecreasing sequence and $\lim\limits_{n \to \infty} x_n = x$ then $x_n \preceq x$ for all $n \in \mathbb{N}$,*
(bii)   *if $\{y_n\}$ is a nondecreasing sequence and $\lim\limits_{n \to \infty} y_n = y$ then $y_n \succeq y$ for all $n \in \mathbb{N}$,*
(biii) *if $\{z_n\}$ is a nondecreasing sequence and $\lim\limits_{n \to \infty} z_n = y$ then $z_n \preceq z$ for all $n \in \mathbb{N}$,*

then F has a tripled fixed point provided that there exist $x_0, y_0, z_0 \in X$ such that

$$x_0 \preceq F(x_0, y_0, z_0), y_0 \succeq F(y_0, x_0, y_0), z_0 \preceq F(z_0, y_0, x_0).$$

Proof. Let x_0, y_0, $z_0 \in X$ be such that

$$x_0 \preceq F(x_0, y_0, z_0), y_0 \succeq F(y_0, x_0, y_0), z_0 \preceq F(z_0, y_0, x_0).$$

As $F(X \times X \times X) \subseteq X$, so we can construct sequences $\{x_n\}, \{y_n\}$, and $\{z_n\}$ in X such that

$$x_{n+1} = F(x_n, y_n, z_n), \ y_{n+1} = F(y_n, x_n, y_n),$$
$$z_{n+1} = F(z_n, y_n, x_n), \forall n \geq 0. \tag{5.1.2}$$

Now we show that

$$x_n \preceq x_{n+1}, y_n \succeq y_{n+1}, z_n \preceq z_{n+1}, \forall n \geq 0. \tag{5.1.3}$$

Since

$$x_0 \preceq F(x_0, y_0, z_0), \ y_0 \succeq F(y_0, x_0, y_0), \ z_0 \preceq F(z_0, y_0, x_0),$$

equation (5.1.3) holds for $n = 0$. Suppose that (5.1.3) holds for any $n \geq 0$. That is,

$$x_n \preceq x_{n+1}, y_n \succeq y_{n+1}, z_n \preceq z_{n+1}. \tag{5.1.4}$$

As F has the *mixed monotone property*, so by (5.1.4) we obtain

$$\begin{cases} F(x_n, y, z) \preceq F(x_{n+1}, y, z) & \text{(i)} \\ F(x, y_n, z) \preceq F(x, y_{n+1}, z) & \text{(ii)} \\ F(x, y, z_n) \preceq F(x, y, z_{n+1}) & \text{(iii)} \end{cases}$$

which on replacing y by y_n and z by z_n in (i) implies that $F(x_n, y_n, z_n) \preceq F(x_{n+1}, y_n, z_n)$, replacing x by x_{n+1} and z by z_n in (ii), we obtain $F(x_{n+1}, y_n, z_n) \preceq F(x_{n+1}, y_{n+1}, z_n)$, replacing y by y_{n+1} and x by x_{n+1} in (iii), we get $F(x_{n+1}, y_{n+1}, z_n) \preceq F(x_{n+1}, y_{n+1}, z_{n+1})$. Thus, we have $F(x_n, y_n, z_n) \preceq F(x_{n+1}, y_{n+1}, z_{n+1})$, that is, $x_{n+1} \preceq x_{n+2}$. Similarly, we have

$$\begin{cases} F(y, x, y_{n+1}) \preceq F(y, x, y_n) & \text{(iv)} \\ F(y_{n+1}, x, y) \preceq F(y_n, x, y) & \text{(v)} \\ F(y, x_{n+1}, y) \preceq F(y, x_n, y) & \text{(vi)} \end{cases}$$

which on replacing y by y_{n+1} and x by x_{n+1} in (iv) implies that $F(y_{n+1}, x_{n+1}, y_{n+1}) \preceq F(y_{n+1}, x_{n+1}, y_n)$, replacing x by x_{n+1} and y by y_{n+1} in (v), we obtain $F(y_{n+1}, x_{n+1}, y_n) \preceq F(y_n, x_{n+1}, y_n)$, replacing y by y_n in (vi), we get $F(y_n, x_{n+1}, y_n) \preceq F(y_n, x_n, y_n)$. Thus, we have $F(y_{n+1}, x_{n+1}, y_{n+1}) \preceq F(y_n, x_n, y_n)$, that is, $y_{n+2} \preceq y_{n+1}$. Similarly, we have

$$\begin{cases} F(z_n, y, x) \preceq F(z_{n+1}, y, x) & \text{(vii)} \\ F(z, y_n, x) \preceq F(z, y_{n+1}, x) & \text{(viii)} \\ F(z, y, x_n) \preceq F(z, y, x_{n+1}) & \text{(ix)} \end{cases}$$

which on replacing y by y_n and x by x_n in (vii) implies that $F(z_n, y_n, x_n) \preceq F(z_{n+1}, y_n, x_n)$, replacing x by x_n and z by z_{n+1} in (viii), we obtain $F(z_{n+1}, y_n, x_n) \preceq F(z_{n+1}, y_{n+1}, x_n)$, replacing y by y_{n+1} and z by z_{n+1} in (ix), we get $F(z_{n+1}, y_{n+1}, x_n) \preceq F(z_{n+1}, y_{n+1}, x_{n+1})$. Thus, we have $F(z_n, y_n, x_n) \preceq F(z_{n+1}, y_{n+1}, x_{n+1})$, that is, $z_{n+1} \preceq z_{n+2}$. So by induction, we conclude that (5.1.4) holds for all $n \geq 0$, that is

$$x_0 \preceq x_1 \preceq x_2 \preceq \ldots \preceq x_n \preceq x_{n+1} \ldots \tag{5.1.5}$$

$$y_0 \succeq y_1 \succeq y_2 \succeq \ldots \succeq y_n \succeq y_{n+1} \ldots \tag{5.1.6}$$

$$z_0 \preceq z_1 \preceq z_2 \preceq \ldots \preceq z_n \preceq z_{n+1} \ldots \tag{5.1.7}$$

Consider,

$$
\begin{aligned}
\mu_{x_n - x_{n+1}}(kt) &= \mu_{F(x_{n-1}, y_{n-1}, z_{n-1}) - F(x_n, y_n, z_n)}(kt) \\
&\geq \min(\mu_{x_{n-1} - x_n}(t), \mu_{y_{n-1} - y_n}(t), \mu_{z_{n-1} - z_n}(t)) \\
&= Q_{x_{n-1} - x_n, y_{n-1} - y_n, z_{n-1} - z_n}(t).
\end{aligned}
\tag{5.1.8}
$$

Also,

$$
\begin{aligned}
\mu_{z_n - z_{n+1}}(kt) &= \mu_{F(z_{n-1}, y_{n-1}, x_{n-1}) - F(z_n, y_n, x_n)}(kt) \\
&\geq \min(\mu_{z_{n-1} - z_n}(t), \mu_{y_{n-1} - y_n}(t), \mu_{x_{n-1} - x_n}(t) \\
&= \min(\mu_{x_{n-1} - x_n}(t), \mu_{y_{n-1} - y_n}(t), \mu_{z_{n-1} - z_n}(t) \\
&= Q_{x_{n-1} - x_n, y_{n-1} - y_n, z_{n-1} - z_n}(t).
\end{aligned}
\tag{5.1.9}
$$

Now,

$$
\begin{aligned}
\mu_{y_n - y_{n+1}}(kt) &= \mu_{F(y_{n-1}, x_{n-1}, y_{n-1}) - F(y_n, x_n, y_n)}(kt) \\
&\geq \min(\mu_{y_{n-1} - y_n}(t), \mu_{x_{n-1} - x_n}(t), \mu_{y_{n-1} - y_n}(t) \\
&= \min(\mu_{y_{n-1} - y_n}(t), \mu_{x_{n-1} - x_n}(t), \mu_{y_{n-1} - y_n}(t)) \\
&\geq \min(\mu_{y_{n-1} - y_n}(t), \mu_{x_{n-1} - x_n}(t), \mu_{y_{n-1} - y_n}(t), \\
&\qquad \mu_{z_{n-1} - z_n}(t), \mu_{z_{n-1} - z_n}(t), \mu_{x_{n-1} - x_n}(t)) \\
&\geq Q_{x_{n-1} - x_n, y_{n-1} - y_n, z_{n-1} - z_n}(t).
\end{aligned}
\tag{5.1.10}
$$

By (5.1.8)-(5.1.10), we obtain

$$Q_{x_n - x_{n+1}, y_n - y_{n+1}, z_n - z_{n+1}}(kt) \geq Q_{x_{n-1} - x_n, y_{n-1} - y_n, z_{n-1} - z_n}(t)$$

for $t > 0$. By Lemma 5.1.4, we conclude that $\{x_n\}, \{y_n\}$, and $\{z_n\}$ are Cauchy sequences in X. Since X is complete, there exist x, y, and z, such that $\lim_{n \to \infty} x_n = x$, $\lim_{n \to \infty} y_n = y$, and $\lim_{n \to \infty} z_n = z$. If the assumption (a) holds, then, we have

$$x = \lim_{n \to \infty} x_{n+1} = \lim_{n \to \infty} F(x_n, y_n, z_n)$$
$$= F(\lim_{n \to \infty} x_n, \lim_{n \to \infty} y_n, \lim_{n \to \infty} z_n) = F(x, y, z),$$

$$y = \lim_{n \to \infty} y_{n+1} = \lim_{n \to \infty} F(y_n, x_n, y_n)$$
$$= F(\lim_{n \to \infty} y_n, \lim_{n \to \infty} x_n, \lim_{n \to \infty} y_n) = F(y, x, y),$$

and

$$z = \lim_{n \to \infty} z_{n+1} = \lim_{n \to \infty} F(z_n, y_n, x_n)$$
$$= F(\lim_{n \to \infty} z_n, \lim_{n \to \infty} y_n, \lim_{n \to \infty} x_n) = F(z, y, x).$$

Suppose that assumption (b) holds, then

$$\mu_{x_{n+1} - F(x,y,z)}(kt)$$
$$= \mu_{F(x_n,y_n,z_n) - F(x,y,z)}(kt) \geq Q_{x_n - x, y_n - y, z_n - z}(t)$$

which on taking limit as $n \to \infty$, gives $\mu_{x - F(x,y,z)}(kt) = 1$, $x = F(x, y, z)$.
Also,

$$\mu_{y_{n+1} - F(y,x,y)}(kt)$$
$$= \mu_{F(y_n,x_n,y_n) - F(y,x,y)}(kt) \geq Q_{y_n - y, x_n - x, y_n - y}(t)$$

which on taking limit as $n \to \infty$, implies $\mu_{y - F(y,x,y)}(kt) = 1$, $y = F(y, x, y)$.
Finally, we have

$$\mu_{z_{n+1} - F(z,y,x)}(kt)$$
$$= \mu_{F(z_n,y_n,x_n) - F(z,y,x)}(kt) \geq Q_{z_n - z, y_n - y, x_n - x}(t)$$

which on taking limit as $n \to \infty$, gives $\mu_{z - F(z,y,x)}(kt) = 1$, $z = F(z, y, x)$.

Theorem 5.1.11 *Let* $(X, \mu, \min)$ *be a complete RN-space,* $\preceq$ *be a partial order on* X. *Let* $F : X \times X \times X \longrightarrow X$, *and* $g : X \longrightarrow X$ *be mappings such that* F *has a mixed g-monotone property and*

$$\mu_{F(x,y,z) - F(u,v,w)}(kt) \tag{5.1.11}$$
$$\geq \min(\mu_{gx - gu}(t), \mu_{gy - gv}(t), \mu_{gz - gw}(t))$$

for all those x, y, z, *and* u, v, w *for which* $gx \preceq gu$, $gy \succeq gv$, $gz \preceq gw$, *where* $0 < k < 1$. *Assume that* $g(X)$ *is complete,* $F(X \times X \times X) \subseteq g(X)$ *and* g *is continuous. If either*

(a)   *F is continuous or*
(b)   *X has the following property:*

(bi)   *if $\{x_n\}$ is a nondecreasing sequence and $\lim\limits_{n \longrightarrow \infty} x_n = x$ then $x_n \preceq x$*
   for all $n \in \mathbb{N}$,

(bii)   *if $\{y_n\}$ is a nondecreasing sequence and $\lim\limits_{n \longrightarrow \infty} y_n = y$ then $y_n \succeq y$*
   for all $n \in \mathbb{N}$, and

(biii)   *if $\{z_n\}$ is a nondecreasing sequence and $\lim\limits_{n \longrightarrow \infty} z_n = y$ then $z_n \preceq z$*
   for all $n \in \mathbb{N}$.

Then F has a tripled coincidence point provided that there exist $x_0, y_0, z_0 \in X$ such that

$$g(x_0) \preceq F(x_0, y_0, z_0), g(y_0) \succeq F(y_0, x_0, y_0), g(z_0) \preceq F(z_0, y_0, x_0).$$

Proof. By Lemma 5.1.9, there exists $E \subseteq X$ such that $g : E \longrightarrow X$ is-one-to one and $g(E) = g(X)$. Now define a mapping $\mathcal{A} : g(E) \times g(E) \times g(E) \longrightarrow X$, by

$$\mathcal{A}(gx, gy, gz) = F(x, y, z) \ \forall \ x, y, z \in X. \tag{5.1.12}$$

Since g is one-to-one, so $\mathcal{A}$ is well defined. Now, (5.1.11) and (5.1.12) imply that

$$\mu_{\mathcal{A}(gx,gy,gz)-\mathcal{A}(gu,gv,gw)}(kt) \tag{5.1.13}$$
$$\geq \min(\mu_{gx-gu}(t), \mu_{gy-gv}(t), \mu_{gz-gw}(t))$$

for all $x, y, z, u, v, w \in E$ for which $gx \preceq gu$, $gy \succeq gv$, $gz \preceq gw$. Since F has a *mixed g-monotone property* for all $x, y, z \in X$, so we have

$$\begin{aligned} x_1, x_2 \in X, \ g(x_1) \preceq g(x_2) &\Longrightarrow F(x_1, y, z) \preceq F(x_2, y, z), \\ y_1, y_2 \in X, \ g(y_1) \preceq g(y_2) &\Longrightarrow F(x, y_2, z) \preceq F(x, y_1, z), \\ z_1, z_2 \in X, \ g(z_1) \preceq g(z_2) &\Longrightarrow F(x, y, z_1) \preceq F(x, y, z_2). \end{aligned} \tag{5.1.14}$$

Now, from (5.1.12) and (5.1.14) we have

$$\begin{aligned} x_1, x_2 \in X, \ g(x_1) \preceq g(x_2) &\Longrightarrow \mathcal{A}(gx_1, gy, gz) \preceq \mathcal{A}(gx_2, gy, gz) \\ y_1, y_2 \in X, \ g(y_1) \preceq g(y_2) &\Longrightarrow \mathcal{A}(gx, gy_2, gz) \preceq \mathcal{A}(gx, gy_1, gz) \\ z_1, z_2 \in X, \ g(z_1) \preceq g(z_2) &\Longrightarrow \mathcal{A}(gx, gy, gz_1) \preceq \mathcal{A}(gx, gy, gz_2). \end{aligned} \tag{5.1.15}$$

Hence, $\mathcal{A}$ has a mixed monotone property. Suppose that assumption (a) holds. Since F is continuous, $\mathcal{A}$ is also continuous. By using Theorem 5.1.10, $\mathcal{A}$ has tripled fixed point $(u, v, w) \in g(E) \times g(E) \times g(E)$. If assumption (b) holds then using the definition of $\mathcal{A}$, following similar arguments to those given in Theorem 5.1.10, $\mathcal{A}$ has a tripled fixed point $(u, v, w) \in g(E) \times g(E) \times g(E)$. Finally, we show that F and g have tripled coincidence point. Since $\mathcal{A}$ has tripled fixed point $(u, v, w) \in g(E) \times g(E) \times g(E)$, we get

$$u = \mathcal{A}(u, v, w), v = \mathcal{A}(v, u, v), w = \mathcal{A}(w, u, v). \tag{5.1.16}$$

Hence, there exist $u_1, v_1, w_1 \in X \times X \times X$ such that $gu_1 = u, gv_1 = v$, and $gw_1 = w$. Now, it follows from (5.1.16) that

$$gu_1 = \mathcal{A}(gu_1, gv_1, w) = F(u_1, v_1, w_1),$$
$$gv_1 = \mathcal{A}(gv_1, gu_1, gv_1) = F(v_1, u_1, v_1),$$
$$gw_1 = \mathcal{A}(gw_1, gu_1, gv_1) = F(w_1, v_1, u_1).$$

Thus, $(u_1, v_1, w_1) \in X \times X \times X$ is tripled coincidence point of F and g.

Example 5.1.12 ([1]) Let $X = \mathbb{R}$. Consider Example 5.1.1 such that $\phi : \mathbb{R}^+ \to (0, 1)$ be defined by $\phi(t) = e^{-\frac{1}{t}}$ for all $t \in \mathbb{R}^+$. Then

$$\mu_x(t) = [\phi(t)]^{|x|}$$

for all $x \in X$ and $t > 0$.

If X is endowed with usual order as $x \preceq y \iff x - y \le 0$, then $(X, \preceq)$ is a partially ordered set. Define mappings $F : X \times X \times X \longrightarrow X$, and $g : X \longrightarrow X$ by

$$F(x, y, z) = 2x - 2y + 2z + 1 \text{ and } g(x) = 7x - 1.$$

Obviously, F and g both are onto maps so $F(X \times X \times X) \subseteq g(X)$, also F and g are continuous and F has mixed g-monotone property. Indeed,

$$x_1, x_2 \in X, \ gx_1 \preceq gx_2 \implies 2x_1 - 2y + 2z + 1 \le 2x_2 - 2y + 2z + 1$$
$$\implies F(x_1, y, z) \preceq F(x_2, y, z).$$

Similarly, we can prove that

$$y_1, y_2 \in X, \ g(y_1) \preceq g(y_2) \implies F(x, y_2, z) \preceq F(x, y_1, z)$$

and

$$z_1, z_2 \in X, \ g(z_1) \preceq g(z_2) \implies F(x, y, z_1) \preceq F(x, y, z_2).$$

If, $x_0 = 0, y_0 = \frac{2}{3}, z_0 = 0$, then

$$-1 = g(x_0) \preceq F(x_0, y_0, z_0) = -\frac{1}{3},$$
$$\frac{11}{3} = g(y_0) \succeq F(y_0, x_0, y_0) = \frac{11}{3},$$
$$-1 = g(z_0) \preceq F(z_0, y_0, x_0) = -\frac{1}{3}.$$

So there exist $x_0, y_0, z_0 \in X$ such that

$$g(x_0) \preceq F(x_0, y_0, z_0), g(y_0) \succeq F(y_0, x_0, y_0), g(z_0) \preceq F(z_0, y_0, x_0).$$

Now for all $x, y, z, u, v, w \in X$, for which $gx \preceq gu$, $gy \succeq gv$, $gz \preceq gw$, we have

$$
\begin{aligned}
&\min(\mu_{gx-gu}(t), \mu_{gy-gv}(t), \mu_{gz-gw}(t)) \\
&= \min(\mu_{7(x-u)}(t), \mu_{7(y-v)}(t), \mu_{7(z-w)}(t)) \\
&= \min(\mu_{(x-u)}\left(\frac{t}{7}\right), \mu_{(y-v)}\left(\frac{t}{7}\right), \mu_{(z-w)}\left(\frac{t}{7}\right)) \\
&= \min(\mu_{(x-u)}\left(\frac{t}{7}\right), \mu_{(v-y)}\left(\frac{t}{7}\right), \mu_{((z-w)}\left(\frac{t}{7}\right)) \\
&\leq \mu_{x-u+v-y+z-w}\left(\frac{3t}{7}\right) \\
&= (e^{-\frac{7}{3t}})^{|(x-u+v-y+z-w)|} \\
&= (e^{-\frac{3.5}{3t}})^{|2(x-u+v-y+z-w)|} \\
&= (e^{-\frac{3.5}{3t}})^{|2(x-u)+2(v-y)+2(z-w)|} \\
&= (e^{-\frac{3.5}{3t}})^{|F(x,y,z)-F(u,v,w)|} \\
&= \mu_{F(x,y,z)-F(u,v,w)}(kt)
\end{aligned}
$$

for $k = \frac{3}{3.5} < 1$. Hence there exists $k = \frac{3}{3.5} < 1$ such that

$$
\mu_{F(x,y,z)-F(u,v,w)}(kt) \\
\geq \min(\mu_{gx-gu}(t), \mu_{gy-gv}(t), \mu_{gz-gw}(t))
$$

for all $x, y, z, u, v, w \in X$, for which $gx \preceq gu$, $gy \succeq gv$, $gz \preceq gw$.

Therefore, all the conditions of Theorem 5.1.11 are satisfied. So F and g have tripled coincidence point and here $\left(\frac{2}{5}, \frac{2}{5}, \frac{2}{5}\right)$ is tripled coincidence point of F and g.

5.1.0.1 Application

Here, we study the existence of a unique solution to an initial value problem, as an application to our tripled fixed point theorem.

Consider the initial value problem

$$
x'(\ell) = f(\ell, x(\ell), x(\ell), x(\ell)), \qquad \ell \in I = [0, 1], \qquad x(0) = x_0, \qquad (5.1.17)
$$

where $f : I \times \mathbb{R} \times \mathbb{R} \times \mathbb{R} \to \mathbb{R}$ and $x_0 \in \mathbb{R}$.

An element $(\alpha, \beta, \gamma) \in C(I, \mathbb{R})^3$ is called a tripled initial value problem (5.1.17) if

$$
\alpha'(\ell) \leq f(\ell, \alpha(\ell), \beta(\ell), \gamma(\ell)),
$$

$\beta'(\ell) \geq f(\ell, \beta(\ell), \alpha(\ell), \beta(\ell))$,

$\delta'(\ell) \leq f(\ell, \gamma(\ell), \beta(\ell), \alpha(\ell))$,

for each $\ell \in I$ together with the initial condition

$$\alpha(0) = \beta(0) = \gamma(0) = x_0.$$

Theorem 5.1.13 *Let* $(C(I, \mathbb{R}), \mu, \min)$ *be a complete RN-space with the following order relation on* $C(I, \mathbb{R})$

$$x, y \in C(I, \mathbb{R}), x \leq y \Longleftrightarrow x(\ell) \leq y(\ell), \quad \forall \ell \in [0, 1],$$

and fuzzy norm

$$\mu_{x-y}(t) = \inf_{\ell \in I} \frac{t}{t + |x(\ell) - y(\ell)|}, \quad x, y \in C(I, \mathbb{R}), \ t > 0.$$

Consider the initial value problem (5.1.17) *with* $f \in C(I \times \mathbb{R}^3, \mathbb{R})$ *which is nondecreasing in the second and fourth variables and nonincreasing in third variable. Suppose that for* $x \geq u$, $y \leq v$, *and* $z \geq w$, *we have*

$$0 \leq f(\ell, x, y, z) - f(\ell, u, v, w) \leq k[(x - u) + (v - y) + (z - w)],$$

where $k \in \left(0, \frac{1}{3}\right)$. *Then the existence of a tripled solution for* (5.1.17) *provides the existence of a unique solution of* (5.1.17) *in* $C(I, \mathbb{R})$.

Proof. The initial value problem (5.1.17) is equivalent to the integral equation

$$x(\ell) = x_0 + \int_0^\ell f(s, x(s), x(s), x(s)) ds, \quad \ell \in I. \tag{5.1.18}$$

Suppose $\{x_n\}$ is a nondecreasing sequence in $C(I, \mathbb{R})$ that converges to $x \in C(I, \mathbb{R})$. Then for every $\ell \in I$, the sequence of the real numbers

$$x_1(\ell) \leq x_2(\ell) \leq \cdots \leq x_n(\ell) \leq \cdots$$

converges to $x(\ell)$. Therefore, for all $\ell \in I$ and $n \in \mathbb{N}$, we have $x_n(\ell) \leq x(\ell)$. Hence, $x_n \leq x$ for all $n \in \mathbb{N}$. Also, $C(I, \mathbb{R}) \times C(I, \mathbb{R}) \times C(I, \mathbb{R})$ is a partially ordered set if we define the following order relation in $X \times X \times X$:

$$(x, y, z) \leq (u, v, w) \Longleftrightarrow x(\ell) \leq u(\ell), \ v(\ell) \leq y(\ell), \text{ and } z(\ell) \leq w(\ell), \ \forall \ell \in I.$$

Define $F : C(I, \mathbb{R}) \times C(I, \mathbb{R}) \times C(I, \mathbb{R}) \longrightarrow C(I, \mathbb{R})$ by

$$F(x, y, z)(\ell) = x_0 + \int_0^\ell f(s, x(s), y(s), z(s)) ds, \quad \ell \in I.$$

Now, for $u \leq x$, $y \leq v$ and $w \leq z$, we have

$$\mu_{F(x,y,z)-F(u,v,w)}(t)$$

$$= \inf_{\ell \in I} \frac{t}{t + \int_0^\ell [f(s, x(s), y(s), z(s)) - f(s, u(s), v(s), w(s))]ds}$$

$$\geq \inf_{\ell \in I} \frac{t}{t + \int_0^\ell k[(x - u) + (v - y) + (z - w)]ds}$$

$$\geq \inf_{\ell \in I} \mathcal{M} \left(\frac{\frac{t}{3}}{\frac{t}{3} + \int_0^\ell k(x - u)ds}, \frac{\frac{t}{3}}{\frac{t}{3} + \int_0^\ell k(v - y)ds}, \frac{\frac{t}{3}}{\frac{t}{3} + \int_0^\ell k(z - w)ds} \right)$$

$$= \min \left(\mu_{x-u}(\frac{t}{3k}), \mu_{y-v}(\frac{t}{3k}), \mu_{z-w}(\frac{t}{3k}) \right)$$

hence

$$\mu_{F(x,y,z)-F(u,v,w)}(3kt) \geq \min \left(\mu_{x-u}(t), \mu_{y-v}(t), \mu_{z-w}(t) \right).$$

Then F satisfies the condition (5.1.1) of Theorem 5.1.10. Now, let (α, β, γ) be a tripled solution of the initial value problem (5.1.17) then we have $\alpha \leq F(\alpha, \beta, \gamma)$, $F(\beta, \alpha, \beta) \leq \beta$, and $\gamma \leq F(\gamma, \beta, \alpha)$. Then Theorem 5.1.10 gives that F has a unique tripled fixed point.

References

1. M. Abbas, B. Ali, W. Sintunavarat, P. Kumam, *Tripled fixed point and tripled coincidence point theorems in intuitionistic fuzzy normed spaces*, Fixed Point Theory Appl., **2012** (2012), 16 pages.
2. R. P. Agarwal, Y. J. Cho, R. Saadati, *On Random topological structures*, Abstr. Appl. Anal., **2011** (2011), 41 pages.
3. A. Ben Amar, S. Chouayekh, A. Jeribi, *New fixed point theorems in Banach algebras under weak topology features and applications to nonlinear integral equations*, J. Funct. Anal., **259** (2010), 2215–2237.
4. C. Alsina, *On the stability of a functional equation arising in probabilistic normed spaces*, in: General Inequalities, vol. 5, Oberwolfach, 1986, Birkhäuser, Basel, (1987), 263–271.
5. C. Alsina, B. Schweizer, A. Sklar, *On the definition of a probabilistic normed space*, Aequat. Math., **46** (1993), 91–98.
6. C. Alsina, B. Schweizer, A. Sklar, *Continuity properties of probabilistic norms*, J. Math. Anal. Appl., **208** (1997), 446–452.
7. T. M. Apostol, *Mathematical Analysis*, Massachusetts: 2nd Edition, Addison-Wesley, (1975).
8. B. Boyd, J. S. W. Wong, *On nonlinear contractions*, Proc. Amer. Math. Soc., **20** (1969), 456–464.
9. V. Berinde, M. Borcut, *Tripled fixed point theorems for contractive type mappings in partially ordered metric spaces*, Nonlinear Anal., **74** (2011), 4889–4897.
10. T. A. Burton, *A fixed point theorem of Krasnoselskii*, Appl. Math. Lett., **11** (1998), 85–88.
11. B. C. Dhage, *On some variants of Schauder's fixed point principle and applications to nonlinear integral equations*, J. Math. Phys. Sci., **25** (1988), 603–611.
12. B. C. Dhage, *On a-condensing mappings in Banach algebras*, Math. Student, **63** (1994), 146–152.
13. B. C. Dhage, *Remarks on two fixed point theorems involving the sum and the product of two operators*, Comput. Math. Appl., **46** (2003), 1779–1785.
14. S. S. Chang, Y. J. Cho, S. M. Kang, *Nonlinear Operator Theory in Probabilistic Metric Spaces*, Nova Science Publishers Inc., New York, (2001).
15. Y. J. Cho, Th. M. Rassias, R. Saadati, *Stability of functional equation in random normed spaces*, Springer, New York, (2014).
16. I. Goleţ, *Some remarks on functions with values in probabilistic normed spaces*, Math. Slovaca, **57** (2007), 259–270.
17. V. Gregori, S. Romaguera, *Some properties of fuzzy metric spaces*, Fuzzy Sets and Systems, **115** (2000), 485–489.

18. O. Hadžić, E. Pap, *Fixed Point Theory in PM-Spaces*, Kluwer Academic Publishers, Dordrecht, (2001).

19. O. Hadžić, E. Pap, *New classes of probabilistic contractions and applications to random operators*, in: Y. J. Cho, J. K. Kim, S. M. Kong (Eds.), Fixed Point Theory and Applications, vol. 4, Nova Science Publishers, Hauppauge, New York, (2003), 97–119.

20. O. Hadžić, E. Pap, M. Budincević, *Countable extension of triangular norms and their applications to the fixed point theory in probabilistic metric spaces*, Kybernetica, **38** (2002), 363–381.

21. O. Hadžić, E. Pap, V. Radu, *Generalized contraction mapping principles in probabilistic metric spaces*, Acta Math. Hungar., **101** (2003), 131–148.

22. R. H. Haghi, Sh. Rezapour, N. Shahzad, *Some fixed point generalizations are not real generalizations*, Nonlinear Anal., **74** (2011), 1799–1803.

23. P. Hajek, *Metamathematics of Fuzzy Logic*, Kluwer Academic Publishers, Dordrecht, (1998).

24. E. P. Klement, R. Mesiar, E. Pap, *Triangular Norms*, Kluwer Academic Publishers, Dordrecht, (2000).

25. E. P. Klement, R. Mesiar, E. Pap, *Triangular norms, Position paper I: basic analytical and algebraic properties*, Fuzzy Sets Syst., **143** (2004), 5–26.

26. E. P. Klement, R. Mesiar, E. Pap, *Triangular norms, Position paper II: general constructions and parameterized amilies*, Fuzzy Sets Syst., **145** (2004), 411–438.

27. E. P. Klement, R. Mesiar, E. Pap, *Triangular norms, Position paper III: continuous t-norms*, Fuzzy Sets Syst., **145** (2004), 439–454.

28. M. A. Krasnoselskii, *Some problems of nonlinear analysis*, Amer. Math. Soc. Trans., **10** (1958), 345–409.

29. B. Lafuerza-Guillén, *D-bounded sets in probabilistic normed spaces and their products*, Rend. Mat., Serie VII, **21** (2001), 17–28.

30. B. Lafuerza-Guillén, *Finite products of probabilistic normed spaces*, Rad. Mat., **13** (2004), 111–117.

31. B. Lafuerza-Guillén, A. Rodríguez-Lallena, C. Sempi, *A study of boundedness in probabilistic normed spaces*, J. Math. Anal. Appl., **232** (1999), 183–196.

32. B. Lafuerza-Guillén, A. Rodríguez-Lallena, C. Sempi, *Normability of probabilistic normed spaces*, Note Mat., **29** (2009), 99–111.

33. R. E. Megginson, *An Introduction to Banach Space Theory*, Springer-Verlag, New York, (1998).

34. K. Menger, *Statistical metrics*, Proc. Nat. Acad. Sci. USA, **28** (1942), 535–537.

35. D. Miheţ, *The probabilistic stability for a functional equation in a single variable*, Acta Math. Hungar., **123** (2009), 249–256.

36. D. Miheţ, *The fixed point method for fuzzy stability of the Jensen functional equation*, Fuzzy Sets Syst., **160** (2009), 1663–1667.

37. D. Miheţ, V. Radu, *On the stability of the additive Cauchy functional equation in random normed spaces*, J. Math. Anal. Appl., **343** (2008), 567–572.

38. D. Miheţ, R. Saadati, S. M. Vaezpour, *The stability of the quartic functional equation in random normed spaces*, Acta Appl. Math., **110** (2010), 797–803.

39. D. Miheţ, R. Saadati, S. M. Vaezpour, *The stability of an additive functional equation in Menger probabilistic φ-normed spaces*, Math. Slovak, **61** (2011), 817–826.

40. D. H. Mushtari, *On the linearity of isometric mappings on random normed spaces*, Kazan Gos. Univ. Uchen. Zap., 128 (1968), 86–90.

41. D. O'Regan, R. Saadati, *Nonlinear contraction theorems in probabilistic spaces*, Appl. Math. Comput., **195** (2008), 86–93.

42. V. Radu, *Some remarks on quasi-normed and random normed structures*, Seminar on Probability Theory and Applications (STPA) 159 (2003), West Univ. of Timişoara.

43. R. Saadati, M. Amini, *D-boundedness and D-compactness in finite dimensional probabilistic normed spaces*, Proc. Indian Acad. Sci., Math. Sci., **115** (2005), 483–492.

44. R. Saadati, D. O'Regan, S. M. Vaezpour, J. K. Kim, *Generalized distance and common fixed point theorems in Menger probabilistic metric spaces*, Bull. Iran. Math. Soc., **35** (2009), 97–117.
45. B. Schweizer, A. Sklar, *Probabilistic Metric Spaces*, Elsevier, North Holand, New York, (1983).
46. A. N. Šerstnev, *On the motion of a random normed space*, Dokl. Akad. Nauk SSSR, **149** (1963), 280–283 (English translation in Soviet Math. Dokl. 4 (1963), 388–390).
47. L. A. Zadeh, *Fuzzy sets*, Inform. Control 8 (1965), 338–353.
48. G. Zhang, M. Zhang, *On the normability of generalized Šerstnev PN spaces*, J. Math. Anal. Appl., **340** (2008), 1000–1011.

Index

Printed in the United States
By Bookmasters